From Calculus to Chaos

From Calculus to Chaos

An Introduction to Dynamics

DAVID ACHESON
Jesus College, Oxford

OXFORD NEW YORK MELBOURNE
OXFORD UNIVERSITY PRESS
1997

Oxford University Press, Great Clarendon Street, Oxford OX2 6DP

Oxford New York
Athens Auckland Bangkok Bogota Bombay
Buenos Aires Calcutta Cape Town Dar es Salaam
Delhi Florence Hong Kong Istanbul Karachi
Kuala Lumpur Madras Madrid Melbourne
Mexico City Nairobi Paris Singapore
Taipei Tokyo Toronto Warsaw

and associated companies in
Berlin Ibadan

Oxford is a trade mark of Oxford University Press

Published in the United States
by Oxford University Press Inc., New York

A catalogue record for this book is available from the British Library

Library of Congress Cataloging in Publication Data
(Data available)

ISBN 0 19 850257 5 (Hbk)
ISBN 0 19 850077 7 (Pbk)

Typeset by Technical Typesetting Ireland
Printed in Great Britain by Bookcraft (Bath) Ltd.,
Midsomer Norton, Avon

Preface

This book is an introduction to some of the most interesting applications of calculus, from Newton's time to the present day. These mainly involve questions of dynamics, i.e. of how and why things change with time.

In taking a fresh approach to the whole subject, I hope the book may be suitable for a wide readership, including

- university students of mathematics and science
- readers who are preparing to take a university course
- mathematics and science teachers, in schools and universities
- 'general' readers who refuse to be put off by a few equations.

The only real requirement is some knowledge of elementary calculus.

My main aim is to help people see, and actually *enjoy*, some truly remarkable applications of mathematics, and the best way I know of doing that is to introduce, by means of simple examples, some of the most exciting results and discoveries, in such a way that the really big ideas do not get obscured in a snowstorm of detail. To this end, we move along quite rapidly, from first steps to the frontiers.

Another novel feature of the book is its approach to the use of computers. It is perfectly possible, of course, to just take the consequent 'results' on trust, but readers who would like to use a PC to explore dynamics for themselves can do so very easily, *even if they have no previous computing experience*, because the book contains an unusually down-to-earth introduction to the whole matter.

While writing the book I have had a great deal of practical advice from potential readers, including teachers and students at both schools and universities. I would like to thank, in particular, Julian Addison, Ian Aitchison, Arthur Barnes, Andrew Bassom, Peter Clifford, Stephen Cox, David Crighton, Tom Evans, John Gittins, Sarah Hennell, Raymond Hide, David Hughes, Mark Mathieson, Janet Mills, Tom Mullin, Paul Newton, Howell Peregrine, John Roe, Helen Sansom and Dan Waterhouse for their helpful comments on individual chapters or the book as a whole. I am also most grateful to those students at Jesus College and Keble College who tried out the various drafts and offered so much advice and encouragement.

Oxford
March 1997

David Acheson

Contents

1 *Introduction*

1.1 The beginning of dynamics

In August 1684 the astronomer Edmund Halley travelled to Cambridge, anxious to seek Newton's advice on the key scientific problem of the day, which concerned planetary motion.

It was already known that each planet moves around the Sun in an ellipse. Moreover, other evidence suggested that the Sun exerts a gravitational force on each planet proportional to $1/r^2$, where r denotes the distance from the Sun. The outstanding question, then, was whether such an inverse-square law of gravitational attraction could account for the elliptical shape of each orbit.

One of Dr. Halley's contemporaries reported of the meeting with Newton that

> ...after they had been some time together, the D^r^ asked him what he thought the Curve would be that would be described by the Planets supposing the force of attraction towards the Sun to be reciprocal to the square of their distance from it. Sr Isaac replied immediately that it would be an Ellipsis, the Doctor struck with joy & amazement asked him how he knew it, Why saith he I have calculated it...

In due course, this single achievement came to be recognized as one of the greatest scientific advances of all time.

Before proceeding further, let us note how dynamical problems are typically solved today, albeit by considering a much simpler example.

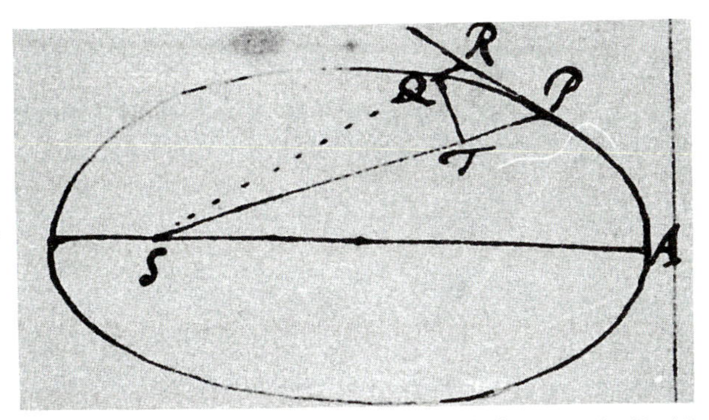

Fig. 1.1 A sketch of orbital motion from Newton's unpublished manuscript De Motu corporum in gyrum *(1684). The moving point P denotes a planet, and the fixed point S denotes the Sun. (Cambridge University Library).*

Fig. 1.2 Isaac Newton (1642–1727).

Suppose that a ball of mass m is initially at a height h above the ground and is then projected horizontally with speed V, as in Fig. 1.3. Gravity will deflect it from a straight-line path, and our problem is to find the subsequent motion.

We first set up a set of **coordinates** (x, y) and view the fundamental physical law

$$\text{force} = \text{mass} \times \text{acceleration} \tag{1.1}$$

as having **components** in the x and y directions. In this case the force components are 0 and $-mg$, if we neglect air resistance, while the components of acceleration are $\mathrm{d}^2x/\mathrm{d}t^2$ and $\mathrm{d}^2y/\mathrm{d}t^2$. In this way we obtain

$$m\frac{\mathrm{d}^2x}{\mathrm{d}t^2} = 0, \qquad m\frac{\mathrm{d}^2y}{\mathrm{d}t^2} = -mg \tag{1.2a,b}$$

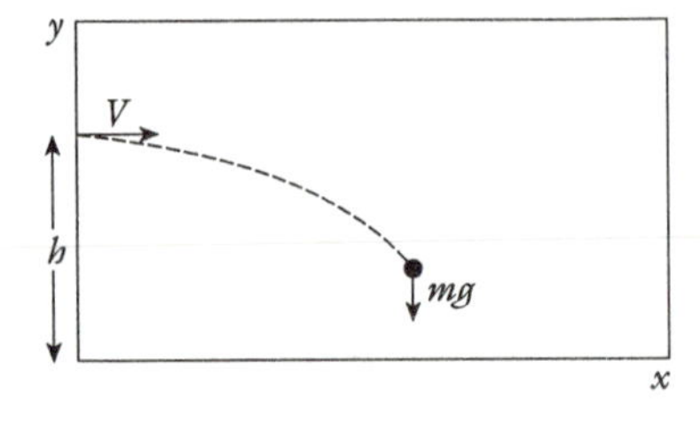

Fig. 1.3 A simple projectile problem.

as our **differential equations of motion.**

We now try to solve these subject to the **initial conditions**, which in this particular case are

$$\left.\begin{aligned} x &= 0, & y &= h \\ \frac{\mathrm{d}x}{\mathrm{d}t} &= V, & \frac{\mathrm{d}y}{\mathrm{d}t} &= 0 \end{aligned}\right\} \quad \text{when } t = 0. \tag{1.3}$$

This whole problem is, fortunately, quite simple. We may at once integrate (1.2b) to obtain

$$\frac{\mathrm{d}y}{\mathrm{d}t} = -gt + c,$$

where c is an arbitrary constant. But one of the initial conditions is that $\mathrm{d}y/\mathrm{d}t = 0$ when $t = 0$, so c must be zero. Integrating again,

$$y = -\tfrac{1}{2}gt^2 + d,$$

and the new constant d must be equal to h in order to satisfy another of the initial conditions, that $y = h$ when $t = 0$. Thus

$$y = h - \tfrac{1}{2}gt^2. \tag{1.4}$$

In a similar way we find by integrating (1.2a) that

$$x = Vt, \tag{1.5}$$

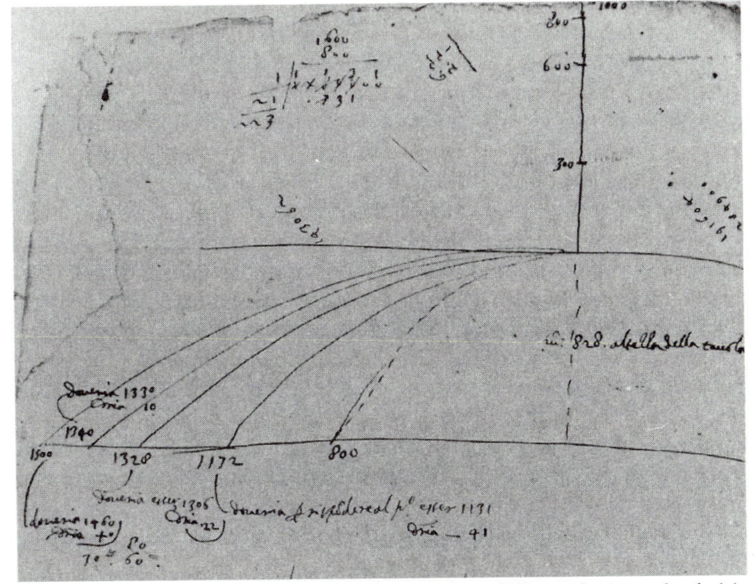

Fig. 1.4 Parabolic paths taken by balls released at various different horizontal velocities V, as measured by Galileo in around 1609. This fragment comes from folio 116v. of his working papers on motion, which are now kept in the Biblioteca Nazionale Centrale, Florence.

Fig. 1.5 Leonhard Euler (1707–1783).

and on eliminating t between (1.4) and (1.5) we find that the path of the ball is

$$y = h - \tfrac{1}{2} g \frac{x^2}{V^2}, \tag{1.6}$$

which is a parabola, as Galileo found by experiment in about 1609 (Fig. 1.4).

Now, Newton's *Principia* of 1687 is justly acclaimed as a great masterpiece of science, but anyone who has actually opened it knows that there is no trace there of any differential equations such as (1.2a, b), nor even of the rest of the approach outlined above. The calculus itself was still in a very early stage of development, and Newton's whole approach to dynamics rests on inspired but *ad hoc* adaptations of the methods of classical geometry (Fig. 1.1).

If we ask, then, where dynamics was first done in the way outlined above, the answer is to be found some 60 years later, in various papers by Leonhard Euler (1707–1783).

Euler was the foremost mathematician of the eighteenth century, and was enormously prolific, producing some 800 scientific papers and several highly original textbooks. And in one of his papers of 1749 we find for the first time not only (1.2) but the general equations of motion for a single point mass m moving in three dimensions:

$$m\frac{d^2x}{dt^2} = F_x, \qquad m\frac{d^2y}{dt^2} = F_y, \qquad m\frac{d^2z}{dt^2} = F_z, \tag{1.7}$$

Cela posé, prenant l'element du tems dt pour conſtant, le changement inſtantané du mouvement du Corps ſera exprimé par ces trois équations :

$$\text{I. } \frac{2ddx}{dt^2} = \frac{X}{M}; \quad \text{II. } \frac{2ddy}{dt^2} = \frac{Y}{M}; \quad \text{III. } \frac{2ddz}{dt^2} = \frac{Z}{M}$$

d'où l'on pourra tirer pour chaque tems ecoulé t les valeurs x, y, z, & par conſéquent l'endroit où le Corps ſe trouvera. C. Q. F. T.

Fig. 1.6 The first statement of the differential equations of particle motion, by Euler, in the Mémoires de l'Académie des Sciences, Berlin, *1749.*

(Fig. 1.6). Here the force acting on the mass has been split into three perpendicular components F_x, F_y, F_z parallel to the various coordinate axes, and the acceleration of the mass has been treated similarly. However commonplace today, this was a novel (and greatly simplifying) idea at the time.

It was roughly in this way, then, and in continental Europe rather than in England, that by the middle of the eighteenth century it at last became clear that the key to an understanding of much of Nature was going to lie in first formulating, and then trying to solve, **differential equations**.

1.2 From calculus to chaos

In this book we shall sketch some of the main applications of calculus, from Newton's time to the present day. While the organization is roughly chronological, so that more recent developments tend to appear later, our priority has been to make each chapter a reasonably self-contained account of some definite topic.

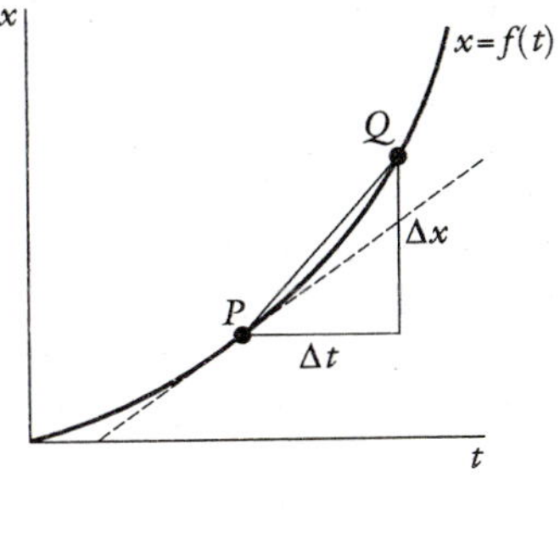

2 *A brief review of calculus*

We begin by highlighting the key results from elementary calculus which we shall need later in the book.

3 *Ordinary differential equations*

We present some simple methods for solving differential equations. This leads us to distinguish between differential equations which are **linear**, like

$$\frac{d^2x}{dt^2} + x = 0,$$

and those which are **nonlinear**. Nonlinear equations are, in general, more

Nonlinear:
$$\frac{d^2x}{dt^2} + x^3 = 0$$

difficult, and their solutions can behave in more diverse and peculiar ways.

4 *Computer solution methods*

The idea here is to convert the differential equation in question into a scheme for calculating a 'new' value of x, at time $t + \Delta t$, in terms of an 'old' value at time t. This up-dating process is then carried out over and over again by computer, using a simple program containing a DO...LOOP. The simplest *step-by-step* method of this kind, due to Euler in 1768, uses the rather crude approximation

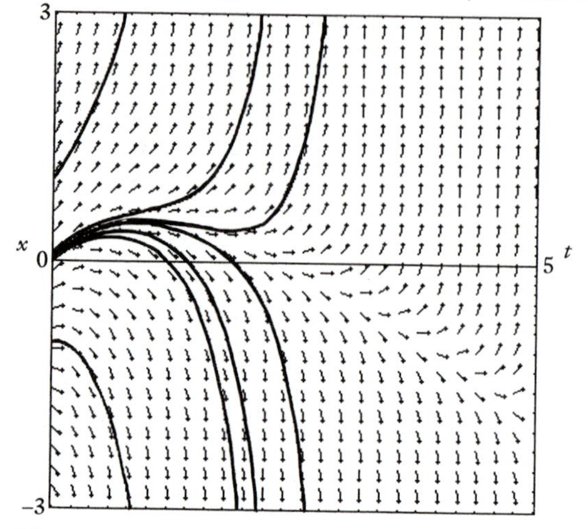

$$\frac{dx}{dt} \doteqdot \frac{x_{new} - x_{old}}{\Delta t}.$$

5 *Elementary oscillations*

Real dynamical problems typically involve non-linear differential equations of second order, but these often simplify greatly if we investigate *small* oscillations about a position of equilibrium. Coupled oscillators are particularly interesting, an early example being the double pendulum, first studied by Euler and Daniel Bernoulli in the 1730s.

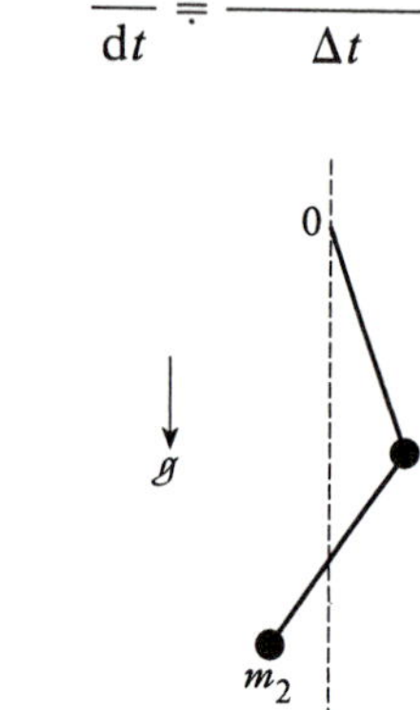

6 *Planetary motion*

We first consider Halley's question to Newton: will a planet orbiting the Sun under an inverse-square law force trace out an ellipse? The answer is yes, but we go on to consider problems involving *three* or more attracting bodies and show that motions of enormous complexity can then occur.

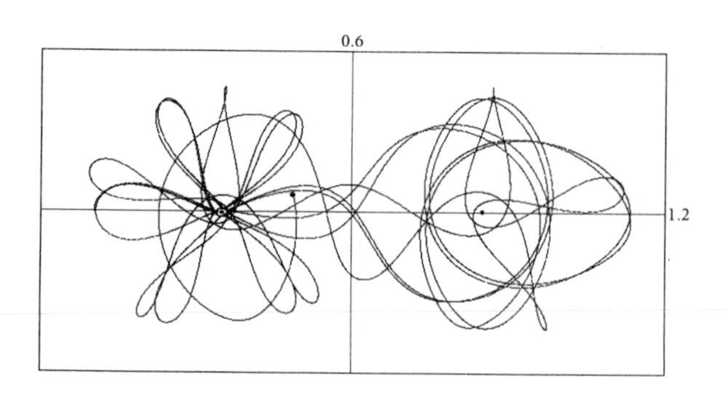

7 *Waves and diffusion*

Many natural phenomena are governed not by so-called ordinary differential equations but by **partial differential equations**. A famous example is the wave equation

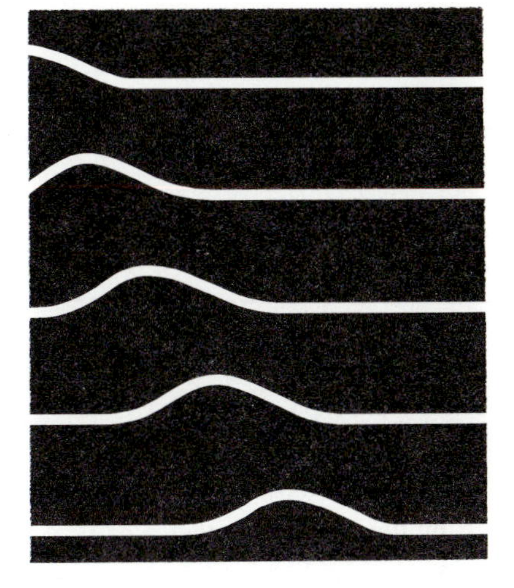

$$\frac{\partial^2 z}{\partial t^2} = c^2 \frac{\partial^2 z}{\partial x^2},$$

with z here being a function of the *two* independent variables x and t. This particular equation arose first in 1745 in connection with vibrating strings, but made a notable reappearance in the 1860s as part of Maxwell's discovery that light is an electromagnetic wave.

8 *The best of all possible worlds?*

Here we look at dynamical problems from a completely different point of view. Instead of thinking in terms of cause and effect, and how a system evolves from time t to time $t + \delta t$, we consider the actual motion *as a whole* and compare it with *other motions that might conceivably have occurred instead.* This approach requires a clever invention called the *calculus of variations*, and is of great value in modern theoretical physics.

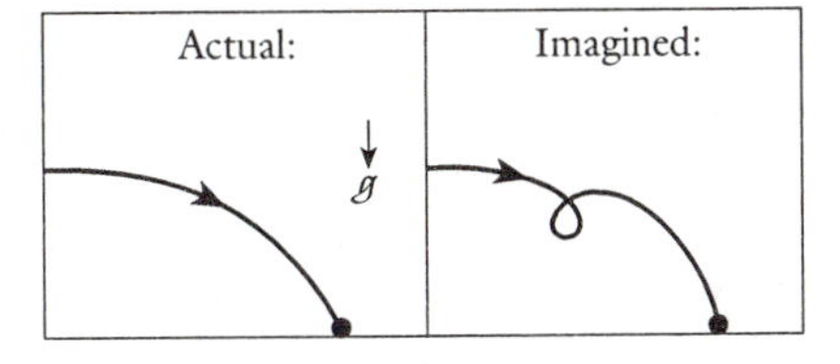

9 *Fluid flow*

Many dynamical problems, such as the airflow past an aeroplane wing, involve fluid motion, and a key quantity in the theory is the *viscosity* of the fluid, denoted by μ. In practice this often seems so small that one is tempted to neglect it altogether, and set $\mu = 0$. While this leads to excellent results in some situations, it leads to complete nonsense in others, *no matter how small the actual value of* μ. The reasons for this paradoxical behaviour began to emerge in 1904, but its consequences are still being investigated today.

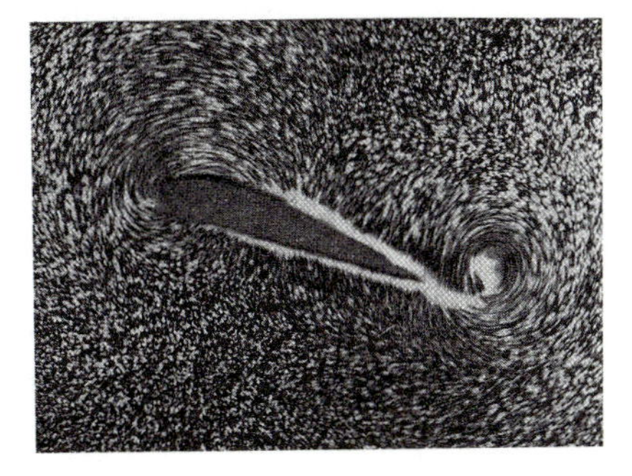

10 *Instability and catastrophe*

It is sometimes possible to infer a great deal about a dynamical system simply by finding its equilibrium states and determining which of these are **stable** to small disturbances and which are **unstable**. This can also help explain sudden or 'catastrophic' jumps from one state to another as some parameter is gradually varied. For example, the clamped elastic strip shown here, which bends under a weight, will *suddenly* spring to a quite different flopped-down state, well to the left of the vertical, as the clamp angle α is gradually increased past a critical value.

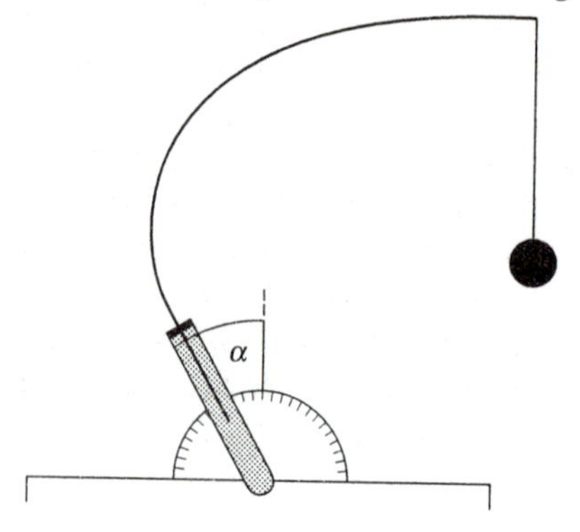

11 *Nonlinear oscillations and chaos*

One of the major discoveries of the last 20 years or so is that even quite 'simple' oscillating systems can display remarkable behaviour if the governing equations are nonlinear. On the one hand, highly regular oscillations can emerge, more or less independently of the initial conditions. At the other extreme, as with the equation

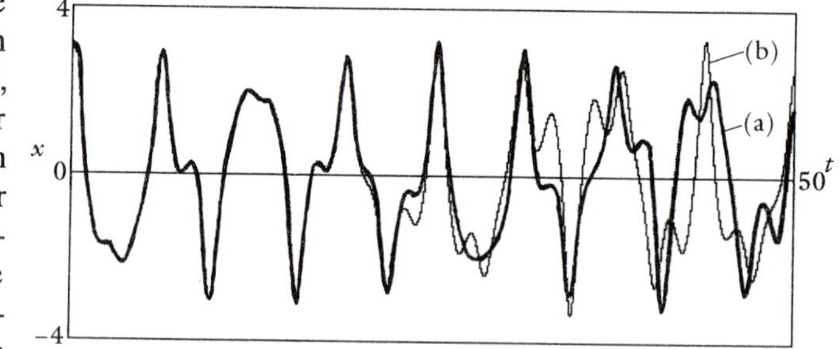

$$\frac{d^2x}{dt^2} + k\frac{dx}{dt} + x^3 = F\cos t,$$

the oscillations can be extremely irregular or **chaotic**, and so sensitive to tiny changes or uncertainties in the initial conditions that long-term predictability of the outcome becomes, in practice, quite impossible.

12 *The not-so-simple pendulum*

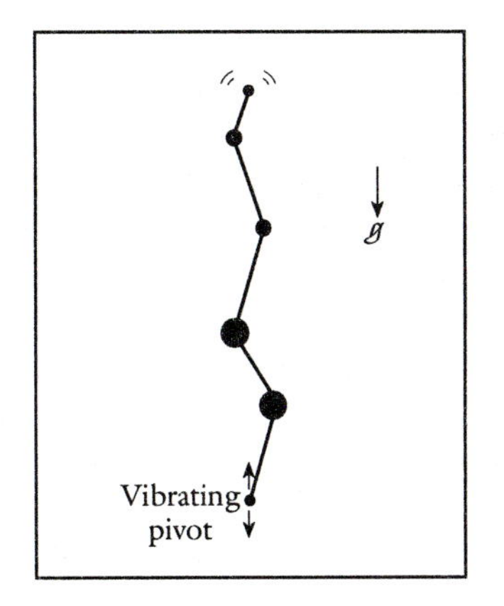

Pendulums have made a strong comeback in the recent scientific literature, mainly because they can provide vivid laboratory demonstrations of chaotic motion. We end the book, however, on a quite different note, with a remarkable balancing act involving an N-pendulum system which has been turned *upside-down.* In this kind of way, apparently, some of the oldest subjects of scientific enquiry are still capable of springing surprises.

2 A brief review of calculus

2.1 Introduction

While calculus is the mathematical key to an understanding of Nature, its roots lie really in problems of geometry.

Thus the **derivative** of a function $y=f(x)$, defined by

$$\frac{dy}{dx}=\lim_{\Delta x\to 0}\frac{f(x+\Delta x)-f(x)}{\Delta x}, \tag{2.1}$$

arises from the problem of finding the tangent to a given curve, and dy/dx represents the *slope* of that tangent, and hence of the curve itself at the point in question (Fig. 2.1). The notation $f'(x)$, rather than dy/dx, is sometimes used.

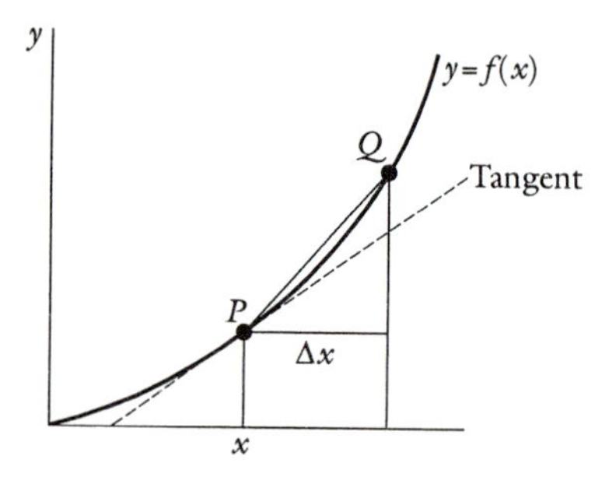

Fig. 2.1 Finding the tangent, by taking Q closer and closer to P.

From the point of view of the present book, **integration** is essentially the opposite process. Thus, given a function $f(x)$, we seek a function $I(x)$ such that

$$\frac{dI}{dx}=f(x), \tag{2.2}$$

and denote the outcome by

$$I=\int f(x)\,dx, \tag{2.3}$$

this **integral** being determined only to within an arbitrary additive constant. Yet there is, again, a geometrical interpretation of integral as the *area under a curve* (Fig. 2.2).

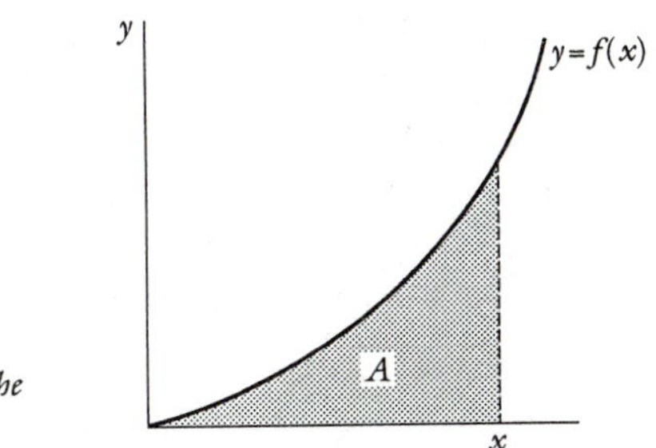

Fig. 2.2 Area under a curve: A has the property $\mathrm{d}A/\mathrm{d}x = f(x)$.

Now, it is not our purpose here to give a systematic account of the calculus; we aim only to collect together, for easy reference, some of the main results, and to recall some of the ways in which these results are related to one another.

2.2 Some elementary results

We assume first that the reader will be familiar with various special cases of differentiation and integration, including

$$\frac{\mathrm{d}}{\mathrm{d}x}(x^n) = nx^{n-1} \tag{2.4}$$

and its counterpart

$$\int x^n \,\mathrm{d}x = \frac{x^{n+1}}{n+1}, \quad n \neq -1. \tag{2.5}$$

Here n need not be an integer; it is not difficult to show that the results hold equally well for *rational* n, i.e. $n = p/q$, where p and q are integers.

Two particularly useful results are

$$\begin{aligned} \frac{\mathrm{d}}{\mathrm{d}x}(\sin x) &= \cos x, \\ \frac{\mathrm{d}}{\mathrm{d}x}(\cos x) &= -\sin x, \end{aligned} \tag{2.6a, b}$$

where radian measure is used.

Fig. 2.3 G. W. Leibniz (1646–1716).

The calculus of more complicated functions is helped by the rules for **differentiating a product or quotient**:

$$\frac{\mathrm{d}}{\mathrm{d}x}(uv) = \frac{\mathrm{d}u}{\mathrm{d}x}\,v + u\,\frac{\mathrm{d}v}{\mathrm{d}x}, \tag{2.7a}$$

$$\frac{\mathrm{d}}{\mathrm{d}x}\left(\frac{u}{v}\right) = \frac{1}{v^2}\left(v\,\frac{\mathrm{d}u}{\mathrm{d}x} - u\,\frac{\mathrm{d}v}{\mathrm{d}x}\right), \tag{2.7b}$$

which Leibniz discovered in 1675. A consequence of the first of these is the rule for **integration by parts**:

$$\int \frac{\mathrm{d}u}{\mathrm{d}x}\,v\,\mathrm{d}x = uv - \int u\,\frac{\mathrm{d}v}{\mathrm{d}x}\,\mathrm{d}x. \tag{2.8}$$

It often happens that we have some variable y which is a function of x, while x itself is some function of another variable, say t. In this way y can be regarded as a function of t, and Leibniz's **chain rule** tells us that

$$\frac{\mathrm{d}y}{\mathrm{d}t} = \frac{\mathrm{d}y}{\mathrm{d}x}\,\frac{\mathrm{d}x}{\mathrm{d}t}. \tag{2.9}$$

This has, again, a natural counterpart in the formula for **integration by substitution**:

$$\int z\,\mathrm{d}x = \int z\,\frac{\mathrm{d}x}{\mathrm{d}t}\,\mathrm{d}t. \tag{2.10}$$

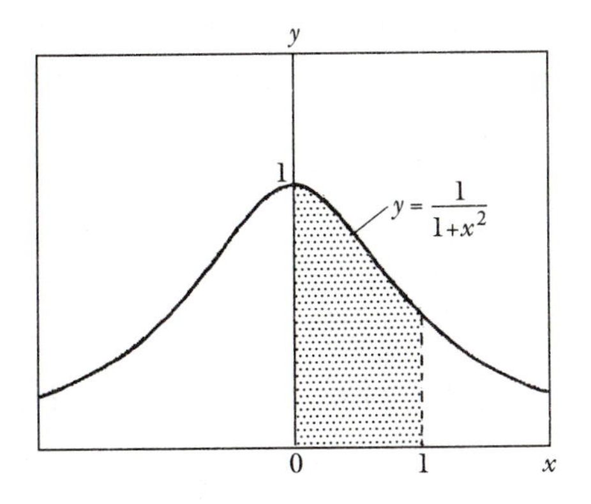

Fig. 2.4 The area corresponding to (2.11).

As an illustration of this last result, consider the area shown in Fig. 2.4, i.e.

$$A = \int_0^1 \frac{1}{1+x^2}\,\mathrm{d}x. \tag{2.11}$$

On introducing the change of variable $x = \tan\theta$ this can be rewritten

$$A = \int_0^{\pi/4} \frac{1}{\sec^2\theta}\frac{\mathrm{d}x}{\mathrm{d}\theta}\,\mathrm{d}\theta = \int_0^{\pi/4} \frac{1}{\sec^2\theta}\cdot\sec^2\theta\,\mathrm{d}\theta = \int_0^{\pi/4}\mathrm{d}\theta,$$

so that we obtain the elegant formula

$$\int_0^1 \frac{1}{1+x^2}\,\mathrm{d}x = \frac{\pi}{4}. \tag{2.12}$$

2.3 Taylor series

We often need an approximation to a function $y = f(x)$ for values of x near to some particular value $x = a$, say, and a crude but obvious way to do this is to draw a straight line through the point in question with the correct slope $f'(a)$, so that

$$y \doteqdot f(a) + (x-a)f'(a) \tag{2.13}$$

(Fig. 2.5).

This takes no account of the local curvature, of course, and we should be able to do rather better by taking a quadratic function of $(x-a)$, i.e. $y = c_0 + c_1(x-a) + c_2(x-a)^2$. We then choose the constants c_0, c_1, c_2 so that the values of y, y' and y'' are all correct at $x = a$; this requires taking $c_0 = f(a)$, $c_1 = f'(a)$, $c_2 = \frac{1}{2}f''(a)$. Continuing in this way we are led to the idea of a **Taylor series** representation for $y = f(x)$ about $x = a$:

$$y = f(a) + (x-a)f'(a) + \frac{(x-a)^2}{2!}f''(a) + \frac{(x-a)^3}{3!}f'''(a) + \cdots. \tag{2.14}$$

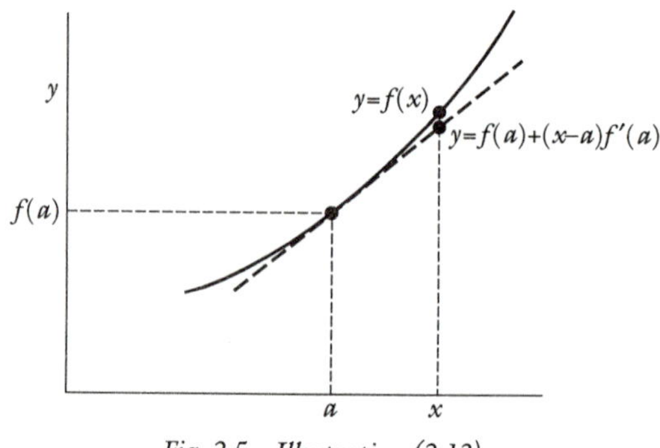

Fig. 2.5 Illustrating (2.13).

Taking $a = 0$, two particular examples are

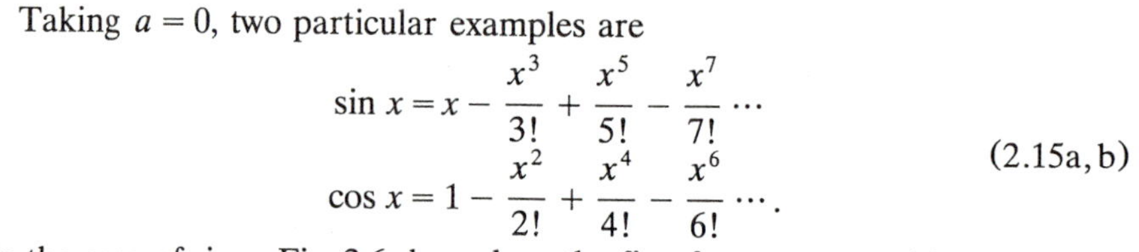

$$\begin{aligned} \sin x &= x - \frac{x^3}{3!} + \frac{x^5}{5!} - \frac{x^7}{7!} \cdots \\ \cos x &= 1 - \frac{x^2}{2!} + \frac{x^4}{4!} - \frac{x^6}{6!} \cdots . \end{aligned} \qquad (2.15\text{a, b})$$

In the case of $\sin x$, Fig. 2.6 shows how the first few terms provide a very good approximation, provided $|x|$ is reasonably small. For larger $|x|$ we need to take more terms to get a good approximation, but the series (2.15a, b) may be used successfully in this way no matter how large $|x|$ is.

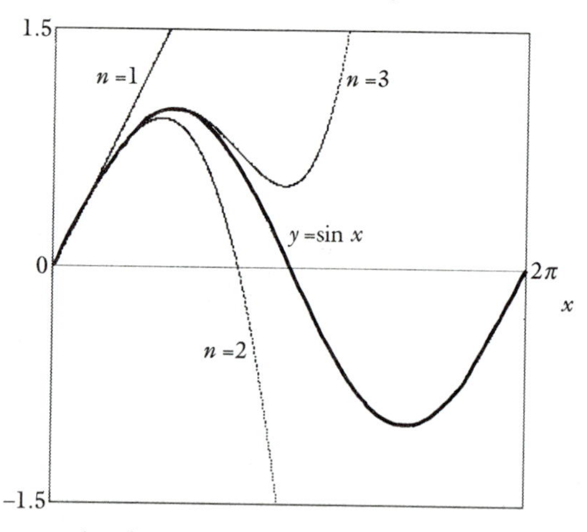

Fig. 2.6 Taylor series approximations to $y = \sin x$, about $x = 0$, using different numbers of terms. Over the interval shown, $n = 8$ gives an extremely good approximation to the actual curve.

More typically, a Taylor series (2.14) converges only for $|x - a|$ less than some definite number R, called the radius of convergence. An important example of this is the **binomial series**

$$(1 + x)^\alpha = 1 + \alpha x + \frac{\alpha(\alpha - 1)}{2!} x^2 + \frac{\alpha(\alpha - 1)(\alpha - 2)}{3!} x^3 + \cdots, \quad (2.16)$$

which converges only for $|x| < 1$. Here α may be any real number, positive or negative, but the importance of the condition $|x| < 1$ can be seen very easily in the particular case $\alpha = -1$. The sum of the first n terms is then

$$1 - x + x^2 \cdots + (-1)^{n-1} x^{n-1} = \frac{1 + (-1)^{n-1} x^n}{1 + x}, \tag{2.17}$$

as we may confirm by multiplying both sides by $1 + x$ and observing all the cancellation. On taking the limit $n \to \infty$ we obtain $(1 + x)^{-1}$, as desired, *provided* $|x| < 1$. This condition is vital, for only then does the term $(-1)^{n-1} x^n$ in the numerator of the right-hand side of (2.17) tend to zero, rather than oscillate ever more wildly as $n \to \infty$.

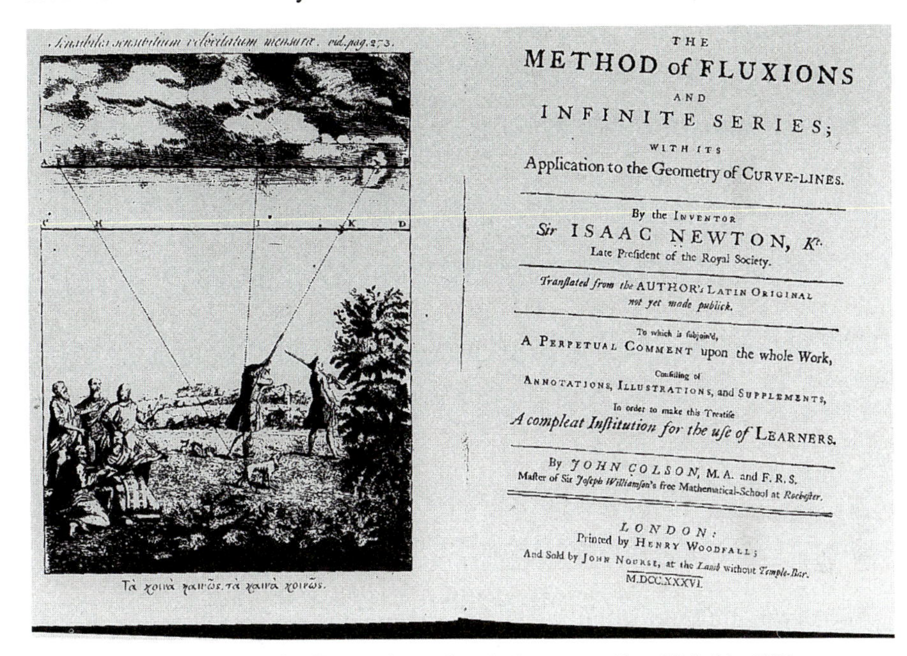

THE
METHOD of FLUXIONS
AND
INFINITE SERIES;
WITH ITS
Application to the Geometry of CURVE-LINES.

By the INVENTOR
Sir ISAAC NEWTON, *K^t.*
Late Preſident of the Royal Society.

Tranſlated from the AUTHOR's LATIN ORIGINAL
not yet made publick.

To which is ſubjoin'd,
A PERPETUAL COMMENT upon the whole Work,
Conſiſting of
ANNOTATIONS, ILLUSTRATIONS, and SUPPLEMENTS,
In order to make this Treatiſe
A compleat Inſtitution for the uſe of LEARNERS.

By *JOHN COLSON*, M.A. and F.R.S.
Maſter of Sir *Joſeph Williamſon's* free Mathematical-School at *Rocheſter.*

LONDON:
Printed by HENRY WOODFALL;
And Sold by JOHN NOURSE, at the *Lamb* without *Temple-Bar.*
M.DCC.XXXVI.

Fig. 2.7 Newton's 1671 treatise on the calculus, eventually published in 1736.

While the main result, (2.14), was published by Brook Taylor in 1715, it was effectively known and used by Newton and others much earlier. Infinite series were, in fact, central to much of Newton's calculus, in the sense that he would often effect an integration by expanding the integrand in an infinite series and then integrate each term separately. If we borrow (2.12), an example of this type of argument is

$$\begin{aligned} \frac{\pi}{4} &= \int_0^1 \frac{1}{1+x^2}\, \mathrm{d}x = \int_0^1 1 - x^2 + x^4 - x^6 \cdots \mathrm{d}x \\ &= \left[x - \frac{x^3}{3} + \frac{x^5}{5} - \frac{x^7}{7} \cdots \right]_0^1, \end{aligned} \tag{2.18}$$

giving the beautiful result

$$\frac{\pi}{4} = 1 - \tfrac{1}{3} + \tfrac{1}{5} - \tfrac{1}{7} + \cdots. \tag{2.19}$$

While this is often credited to Gregory and Leibniz, it was apparently first discovered by Indian mathematicians some 150 years earlier.

2.4 The functions e^x and $\log x$

From the point of view of the present book, the key property of the **exponential** function $y = \exp(x)$ is that

$$\frac{d}{dx}[\exp(x)] = \exp(x), \tag{2.20}$$

so that it is *equal to its own derivative*. We shall take this, together with

$$\exp(0) = 1, \tag{2.21}$$

as our starting point.

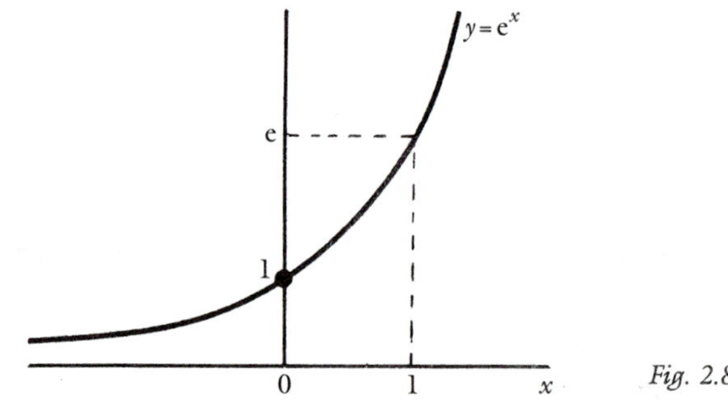

Fig. 2.8 The exponential function.

Successive differentiation of (2.20) shows, in fact, that all the higher derivatives of $\exp(x)$ are also equal to $\exp(x)$, and (2.21) then implies that they are all equal to 1 when $x = 0$. The Taylor series (2.14) for $y = \exp(x)$ about $x = 0$ is therefore

$$\exp(x) = 1 + x + \frac{x^2}{2!} + \frac{x^3}{3!} + \cdots \tag{2.22}$$

and it can be shown that this converges for all x.

If we *define* $e = \exp(1)$, then

$$\begin{aligned} e &= 1 + 1 + \frac{1}{2!} + \frac{1}{3!} + \cdots \\ &= 2.71828\,18284\,59045\ldots. \end{aligned} \tag{2.23}$$

Using (2.20) and (2.21) we may show that

$$\exp(x+y)=\exp(x)\exp(y) \tag{2.24}$$

(Ex. 2.2). It follows that

$$\exp(nx)=[\exp(x)]^n,$$

and, in particular, that

$$\exp(n)=e^n \tag{2.25}$$

for positive integer values of n. The result can in fact be extended without too much difficulty to any *rational* value of n, and as elementary mathematics ascribes no clear meaning to e^x when x is irrational it is only natural to *define* e^x, when x is irrational, as $\exp(x)$. In this way

$$\exp(x)=e^x \tag{2.26}$$

for all real x.

By combining (2.20), (2.26) and the chain rule (2.9) we may show that

$$\frac{d}{dx}(e^{kx})=k\,e^{kx} \tag{2.27}$$

for any constant k, and we shall often use this result.

The function $\log x$

We may define the function log as the inverse of the exponential function, so that

$$z=\log x \quad \Leftrightarrow \quad x=e^z. \tag{2.28}$$

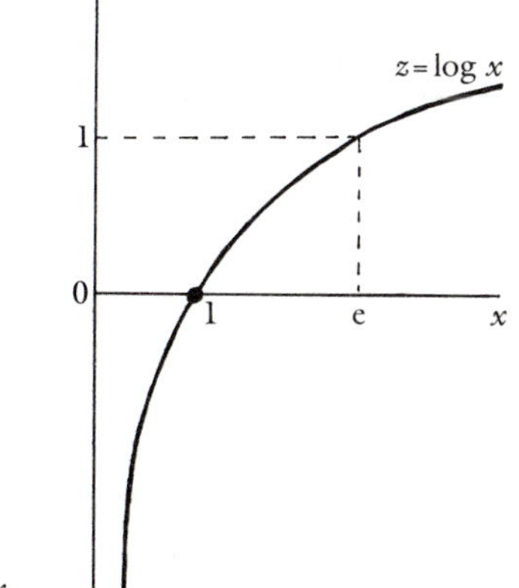

Fig. 2.9 The function $\log x$.

Some important consequences are

$$\log 1 = 0, \tag{2.29}$$

$$\log(ab) = \log a + \log b, \tag{2.30}$$

and it also follows that

$$\frac{\mathrm{d}}{\mathrm{d}x}(\log x) = \frac{1}{x}, \tag{2.31}$$

so that the 'missing' integral in the set (2.5) has now been identified: the integral of x^{-1} is $\log x$ (plus a constant).

Euler's formula for $\mathrm{e}^{\mathrm{i}\theta}$

The Taylor series (2.22) can be shown to be valid for all *real* x. If we are reckless enough to set $x = \mathrm{i}\theta$, where $\mathrm{i} = \sqrt{-1}$ and θ is real, we obtain

$$\begin{aligned}\mathrm{e}^{\mathrm{i}\theta} &= 1 + \mathrm{i}\theta - \frac{\theta^2}{2!} - \frac{\mathrm{i}\theta^3}{3!} + \frac{\theta^4}{4!} + \frac{\mathrm{i}\theta^5}{5!} \cdots \\ &= \left(1 - \frac{\theta^2}{2!} + \frac{\theta^4}{4!} \cdots\right) + \mathrm{i}\left(\theta - \frac{\theta^3}{3!} + \frac{\theta^5}{5!} \cdots\right),\end{aligned} \tag{2.32}$$

and on using (2.15) this becomes

$$\mathrm{e}^{\mathrm{i}\theta} = \cos\theta + \mathrm{i}\sin\theta. \tag{2.33}$$

This extraordinary formula was originally derived by Euler in a quite different way (Fig. 2.10). We must really view it as a *definition* of $\mathrm{e}^{\mathrm{i}\theta}$, for we have not yet established any meaning to a number raised to an imaginary power. Nonetheless, we soon find that $\mathrm{e}^{\mathrm{i}\theta}$ behaves according to all the 'usual' rules. In particular,

$$\frac{\mathrm{d}}{\mathrm{d}\theta}(\mathrm{e}^{\mathrm{i}\theta}) = \mathrm{i}\mathrm{e}^{\mathrm{i}\theta} \tag{2.34}$$

and products behave according to the usual index law

$$\mathrm{e}^{\mathrm{i}\theta_1}\,\mathrm{e}^{\mathrm{i}\theta_2} = \mathrm{e}^{\mathrm{i}(\theta_1+\theta_2)} \tag{2.35}$$

(Ex. 2.5). In consequence,

$$(\mathrm{e}^{\mathrm{i}\theta})^n = \mathrm{e}^{\mathrm{i}n\theta},$$

where n is any positive integer, so

$$(\cos\theta + \mathrm{i}\sin\theta)^n = \cos n\theta + \mathrm{i}\sin n\theta, \tag{2.36}$$

a result known as De Moivre's theorem.

Finally, on setting $\theta = \pi$ in Euler's formula (2.33) we obtain

$$\mathrm{e}^{\mathrm{i}\pi} = -1, \tag{2.37}$$

138. Ponatur denuo in formulis § 133 arcus z infinite parvus et sit n numerus infinite magnus i, ut iz obtineat valorem finitum v. Erit ergo $nz = v$ et $z = \frac{v}{i}$, unde sin. $z = \frac{v}{i}$ et cos. $z = 1$; his substitutis fit

$$\cos. v = \frac{\left(1+\frac{v\sqrt{-1}}{i}\right)^i + \left(1-\frac{v\sqrt{-1}}{i}\right)^i}{2}$$

atque

$$\sin. v = \frac{\left(1+\frac{v\sqrt{-1}}{i}\right)^i - \left(1-\frac{v\sqrt{-1}}{i}\right)^i}{2\sqrt{-1}}$$

In capite autem praecedente vidimus esse

$$\left(1+\frac{z}{i}\right)^i = e^z$$

denotante e basin logarithmorum hyperbolicorum; scripto ergo pro z partim $+v\sqrt{-1}$ partim $-v\sqrt{-1}$ erit

$$\cos. v = \frac{e^{+v\sqrt{-1}} + e^{-v\sqrt{-1}}}{2}$$

et

$$\sin. v = \frac{e^{+v\sqrt{-1}} - e^{-v\sqrt{-1}}}{2\sqrt{-1}}.$$

Ex quibus intelligitur, quomodo quantitates exponentiales imaginariae ad sinus et cosinus arcuum realium reducantur. Erit vero

$$e^{+v\sqrt{-1}} = \cos. v + \sqrt{-1}\cdot\sin. v$$

et

$$e^{-v\sqrt{-1}} = \cos. v - \sqrt{-1}\cdot\sin. v.$$

Fig. 2.10 Euler's formula (2.33), as first stated in his Introductio in analysin infinitorum *(1748). Compare the third displayed expression with (2.38); at this time the concept of a limit had still not been clearly formulated, and i is being used here to denote infinity (∞). Only later did Euler introduce the notation* $i=\sqrt{-1}$.

which relates the three fundamental quantities e, π and i and is widely regarded as one of the most beautiful equations in the whole of mathematics.

Exercises

2.1 *Stationary points.* A function $y=f(x)$ is said to have a stationary point wherever $\mathrm{d}y/\mathrm{d}x=0$. Find all such points in the case

$$y = x^3 - ax + 1,$$

where a is a constant, and examine the sign of $\mathrm{d}^2y/\mathrm{d}x^2$ at each one to determine whether y has a local maximum or minimum there.

2.2 Show directly from (2.20) and (2.21) that

$$\exp(x+y)=\exp(x)\exp(y),$$

by considering

$$\frac{\exp(x+y)}{\exp(x)},$$

with y held constant.

2.3 *Another view of* e^x. The interesting result

$$\lim_{n\to\infty}\left(1+\frac{\alpha}{n}\right)^n=\mathrm{e}^\alpha, \tag{2.38}$$

with α fixed, independent of the integer n, provides a quite different method from (2.23) for the calculation of e, though convergence is extremely slow (Table 2.1).

Table 2.1 Convergence to e using (2.38)

n	$(1+(1/n))^n$
1	2.0000
10	2.5937
100	2.7048
1000	2.7169
10 000	2.7181

Prove (2.38) by first observing that (2.31) implies that

$$\lim_{\Delta x\to 0}\frac{\log(x+\Delta x)-\log x}{\Delta x}=\frac{1}{x},$$

and then setting $\Delta x=1/n$.

2.4 *The functions* $\cosh x$ *and* $\sinh x$. These are defined as follows:

$$\cosh x=\tfrac{1}{2}(\mathrm{e}^x+\mathrm{e}^{-x}),\qquad \sinh x=\tfrac{1}{2}(\mathrm{e}^x-\mathrm{e}^{-x}).$$

Show that

$$\frac{\mathrm{d}}{\mathrm{d}x}\cosh x=\sinh x,\qquad \frac{\mathrm{d}}{\mathrm{d}x}\sinh x=\cosh x$$

and

$$\cosh^2 x-\sinh^2 x=1.$$

2.5 Prove that

$$\mathrm{e}^{\mathrm{i}\theta_1}\,\mathrm{e}^{\mathrm{i}\theta_2}=\mathrm{e}^{\mathrm{i}(\theta_1+\theta_2)}.$$

3 Ordinary differential equations

3.1 Introduction

In this chapter we consider some methods for solving first and second order differential equations.

We begin with first order equations, i.e. equations of the form

$$\frac{dx}{dt} = f(x, t), \tag{3.1}$$

where $f(x, t)$ is some given function of x and t. There will also be an initial condition

$$x = x_0 \quad \text{at} \quad t = 0, \tag{3.2}$$

x_0 being a given constant.

While we focus attention here on trying to find an exact expression for x in terms of t, we shall bear in mind that the question of most practical interest is often: 'what *eventually* happens to x?'. In particular,

(i) does $|x| \to \infty$?
(ii) does x tend to some finite limit?
(iii) does x settle down into some kind of regular oscillation?
(iv) might it perhaps do none of these things, but instead fluctuate erratically (while remaining bounded), even as $t \to \infty$?

(see Fig. 3.1).

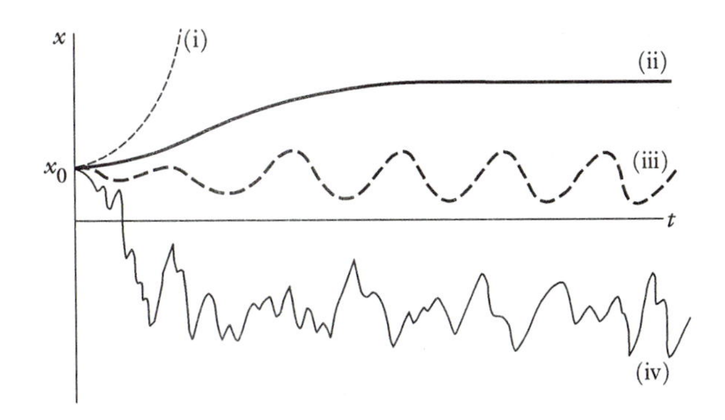

Fig. 3.1 Four conceivable possibilities for the eventual behaviour of the solution $x(t)$ to a differential equation and given initial condition(s).

We shall also emphasize, from the very outset, a *geometrical* view of differential equations. Consider, for example, the equation

$$\frac{dx}{dt} = \lambda x, \tag{3.3}$$

where λ is a constant. This arises, when $\lambda > 0$, from the simplest of population models, namely birth rate dx/dt proportional to population x. Using (2.27) we can readily confirm that

$$x = x_0 e^{\lambda t} \tag{3.4}$$

satisfies (3.3) and the initial condition (3.2), and we have plotted several solution curves, corresponding to different values of x_0, in Fig. 3.2, for the particular case $\lambda = 1$. But we have also used the fact that the differential equation (3.1) *itself* gives the slope $f(x, t)$ of the solution curve passing through any particular point of the (t, x) plane. In Fig. 3.2 we have therefore attached a short line segment of appropriate slope (i.e. x, in this particular case) to each point of a grid in the (t, x) plane, with a crude 'arrow' at the other end. The various solution curves are seen to follow this **direction field** in the obvious way.

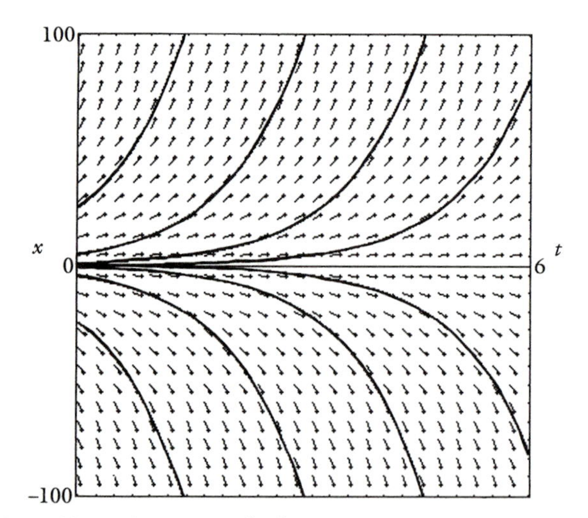

Fig. 3.2 Direction field for the equation (3.3) in the case $\lambda = 1$, and solution curves for $|x_0| = 0.2, 1, 5, 25$.

In practice, the real value of this geometrical approach comes when the original equation (3.1) is so difficult that we are unable to obtain an explicit solution such as (3.4). In those circumstances we may still construct the direction field, and then, given an initial value x_0, we may, in principle, construct the corresponding solution curve by tracing it from the point $(0, x_0)$ in such a way as to follow the direction field at each successive point.

That said, our major concern in this chapter will be identifying special circumstances in which we can solve the differential equation exactly.

Linear versus nonlinear equations

Throughout the book we will need to distinguish clearly between differential equations which are linear and those which are not.

A linear differential equation is one in which the 'unknown' or dependent variable x *and its various derivatives* appear in a linear way. Thus

$$\frac{\mathrm{d}x}{\mathrm{d}t} = c_1 x + c_2,$$

where c_1 and c_2 are constants, is linear, because x and $\mathrm{d}x/\mathrm{d}t$ appear only to the first power. It would still be linear if c_1 and c_2 were complicated functions of t—it is how the *dependent* variable x and its derivatives appear that matters—but

$$\frac{\mathrm{d}x}{\mathrm{d}t} = (1 - x)x$$

is a *non*linear equation, because of the x^2 term, and so are

$$\frac{\mathrm{d}x}{\mathrm{d}t} = \sin x$$

and

$$x\frac{\mathrm{d}x}{\mathrm{d}t} = x + t.$$

The major simplifying feature of linear differential equations becomes really evident only when we come to consider equations of second order, in Section 3.4.

3.2 First-order linear equations

Equation (3.1) is linear if $f(x, t)$ is of the form $a(t)x + b(t)$, and equation (3.3) is of this type, with $a(t) = \lambda$ and $b(t) = 0$. We may tackle any such equation, in principle, by the **method of integrating factors**.

Consider as an example the linear equation

$$\frac{\mathrm{d}x}{\mathrm{d}t} + 2x = t. \tag{3.5}$$

If we multiply both sides by e^{2t} we *turn the left-hand side into the derivative of a product*, for

$$e^{2t}\frac{dx}{dt}+2e^{2t}x=t\,e^{2t}$$

may be rewritten as

$$\frac{d}{dt}(x\,e^{2t})=t\,e^{2t}. \tag{3.6}$$

Integrating, using integration by parts on the right-hand side, we obtain

$$x\,e^{2t}=\tfrac{1}{2}t\,e^{2t}-\tfrac{1}{4}e^{2t}+c.$$

On applying the initial condition (3.2) we find that $c=x_0+\frac{1}{4}$, so

$$x=\tfrac{1}{2}t-\tfrac{1}{4}+(x_0+\tfrac{1}{4})\,e^{-2t}. \tag{3.7}$$

Various solution curves corresponding to different values of x_0 are shown in Fig 3.3, together with the direction field, the slope of each line segment being $t-2x$ in this case, according to (3.5). It is evident both from the figure and from (3.7) that as $t\to\infty$ the solution curve approaches ever more closely the line $x=\frac{1}{2}t-\frac{1}{4}$, regardless of the initial conditions, and that the solution *is* this line in the special case $x_0=-\frac{1}{4}$.

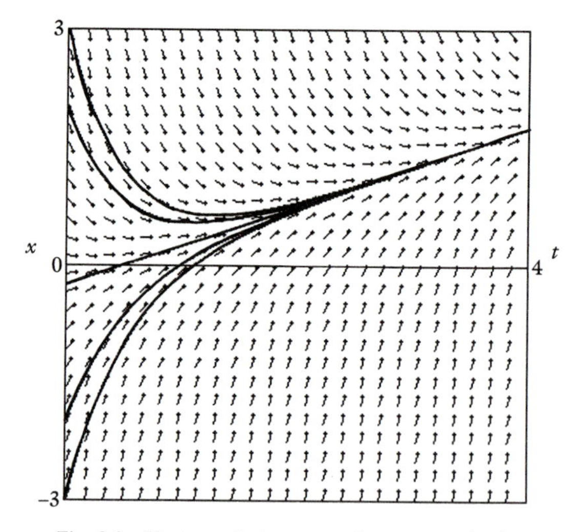

Fig. 3.3 Various solution curves for equation (3.5).

A little reflection shows that we may, in principle, apply the above method to *any* first-order linear equation

$$\frac{dx}{dt}+p(t)x=q(t), \tag{3.8}$$

$p(t)$ and $q(t)$ being given functions of t; the corresponding *integrating factor* by which we first multiply both sides is

$$I = e^{\int p(t)\,dt}, \tag{3.9}$$

for this has the desired property that its derivative with respect to t is $p(t)$ times itself. In the above example, $p(t) = 2$, so $I = e^{2t}$.

3.3 First-order nonlinear equations

We shall divide nonlinear equations in general into two classes, according to whether or not they are *autonomous*.

Autonomous equations

The first-order equation (3.1) is said to be **autonomous** (i.e. 'self-governing') if the rate of change of x is simply a function of x itself, and not dependent explicitly on t:

$$\frac{dx}{dt} = f(x). \tag{3.10}$$

One consequence of (2.9) is that $dx/dt = (dt/dx)^{-1}$, so we may rewrite (3.10) in the form

$$\frac{dt}{dx} = \frac{1}{f(x)}, \tag{3.11}$$

and then integrate with respect to x to obtain

$$t = \int \frac{1}{f(x)}\,dx. \tag{3.12}$$

We consider next a number of examples.

'Blow-up'

The equation

$$\frac{dx}{dt} = x^2 \tag{3.13}$$

is nonlinear, on account of the x^2 term, but of the form (3.10). We find from (3.12) that $t = c - 1/x$, and on applying the initial condition (3.2) we obtain

$$x = \frac{1}{\dfrac{1}{x_0} - t}. \tag{3.14}$$

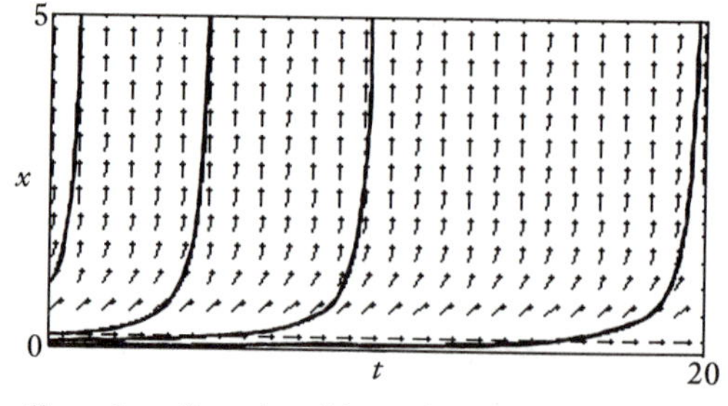

Fig. 3.4 A possible effect of nonlinearity: 'blow-up' in a finite time. The initial values are $x_0 = 1, 0.2, 0.1, 0.05$.

The solution curves for various different (positive) values of x_0 are shown in Fig. 3.4, and the most notable feature is the 'blow-up' of the solution as t approaches $1/x_0$. In sharp contrast to (3.3) or (3.5), then, *the solution exists only for a finite time*. Moreover, the 'blow-up time' is determined by the initial condition; the larger the value of x_0, the sooner the solution breaks down.

An epidemic model

Suppose that a fraction x of a population has an infectious disease, so that a fraction $S = 1 - x$ does not. In the simplest model of the spreading of the disease we assume that members of the population can meet freely, and that the rate of increase of x is then proportional to both x and S, i.e.

$$\frac{\mathrm{d}x}{\mathrm{d}t} = rx(1-x), \tag{3.15}$$

where r is a positive constant.

Then

$$\begin{aligned} rt &= \int \frac{1}{x(1-x)}\,\mathrm{d}x \\ &= \int \frac{1}{x} + \frac{1}{1-x}\,\mathrm{d}x \\ &= \log x - \log(1-x) + c. \end{aligned}$$

It follows that

$$\frac{x}{1-x} = A\,\mathrm{e}^{rt} \tag{3.16}$$

and hence that

$$x = \frac{1}{1 + B\,\mathrm{e}^{-rt}},$$

where $B = A^{-1}$. On applying the initial condition (3.2) we finally obtain

$$x = \frac{1}{1 - \left(1 - \dfrac{1}{x_0}\right) e^{-rt}}. \tag{3.17}$$

The most evident feature of this solution is that $x \to 1$ as $t \to \infty$, so that everyone gets the disease sooner or later, no matter how few are infected initially (unless $x_0 = 0$, in which case $x = 0$ for all t). Happily, this epidemic model is over-simplified, and takes no account, for example, of the possibility that some infected people might be isolated, or even get better.

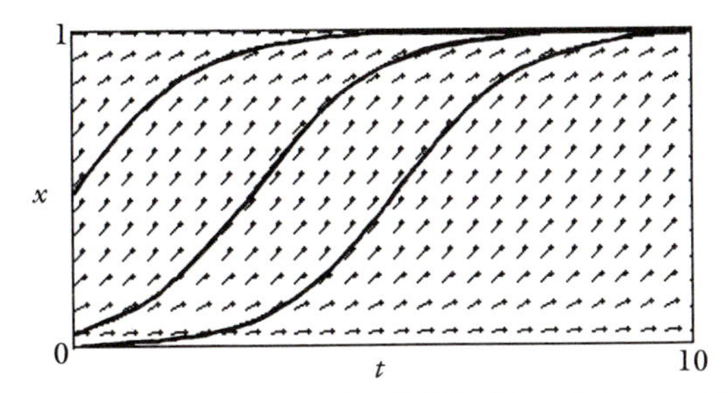

Fig. 3.5 Some solutions to the epidemic equation (3.15), with $r = 1$, for $x_0 = 0.5$, 0.05 and 0.005.

Impossibility of oscillations

An immediate, and rather stark, consequence of the 'geometric' point of view is that no autonomous equation (3.10) can have a solution $x(t)$ which *oscillates*, whether in the manner of (iii) or (iv) in Fig. 3.1. This is true no matter how cleverly we choose the function $f(x)$, and it is true simply because the direction field is independent of t, as in the special cases of Figs. 3.2, 3.4 and 3.5, so that a solution curve which starts by 'going up' can never 'come down' again, and vice versa.

Non-autonomous equations

We may obtain exact solutions to these equations only in certain special circumstances.

One of the most common is when the function $f(x, t)$ in (3.1) is the product of a function of x and a function of t, i.e.

$$\frac{\mathrm{d}x}{\mathrm{d}t} = g(x)h(t), \tag{3.18}$$

and the equation is then said to be **separable**. This is because we may write

$$\frac{1}{g(x)} \frac{\mathrm{d}x}{\mathrm{d}t} = h(t)$$

and then integrate both sides with respect to t to obtain

$$\int \frac{1}{g(x)} \frac{dx}{dt}\, dt = \int h(t)\, dt.$$

In view of (2.10) this may be written

$$\int \frac{1}{g(x)}\, dx = \int h(t)\, dt, \tag{3.19}$$

and provided the functions g and h are simple enough for us to carry out these two integrations we obtain a direct relationship between x and t.

As an example, consider

$$\frac{dx}{dt} = (1-t)x^2, \tag{3.20}$$

which leads to

$$\int \frac{1}{x^2}\, dx = \int (1-t)\, dt,$$

so

$$-\frac{1}{x} = t - \tfrac{1}{2}t^2 + c.$$

On applying the initial condition $x = x_0$ at $t = 0$ we obtain

$$x = \frac{1}{\dfrac{1}{x_0} - \frac{1}{2} + \frac{1}{2}(1-t)^2}. \tag{3.21}$$

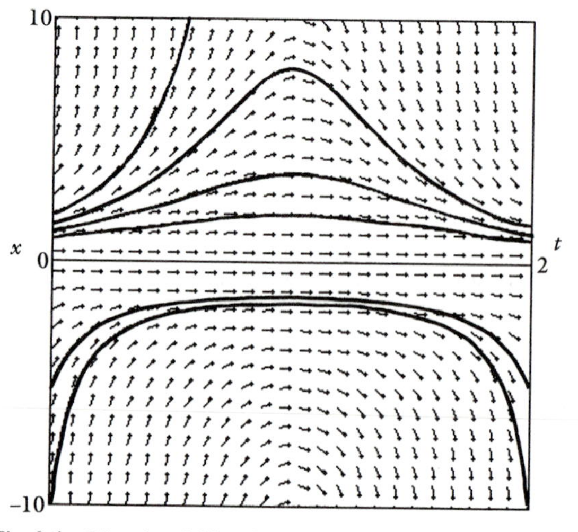

Fig. 3.6 Direction field and various solution curves for (3.20).

3.4 Second-order linear equations

A *linear* ordinary differential equation of second order is of the form

$$a\frac{\mathrm{d}^2x}{\mathrm{d}t^2}+b\frac{\mathrm{d}x}{\mathrm{d}t}+cx=d, \tag{3.22}$$

where a, b, c, d may be constants, or given functions of t, but must be independent of x and its various derivatives.

A major simplifying feature of linear equations becomes evident if we consider the so-called homogeneous case, when $d = 0$.

Homogeneous linear equations

Suppose that we have somehow found two particular solutions to the equation

$$a\frac{\mathrm{d}^2x}{\mathrm{d}t^2}+b\frac{\mathrm{d}x}{\mathrm{d}t}+cx=0, \tag{3.23}$$

We know, in other words, two functions $x = x_1(t)$ and $x = x_2(t)$, say, such that $a\ddot{x}_1 + b\dot{x}_1 + cx_1 = 0$ and $a\ddot{x}_2 + b\dot{x}_2 + cx_2 = 0$, where a dot denotes differentiation with respect to t. Then by substituting directly into the equation (3.23) we may verify that

$$x = Ax_1(t) + Bx_2(t) \tag{3.24}$$

is also a solution, for any values of the constants A and B (Ex. 3.3). Provided that one of the particular solutions $x_1(t), x_2(t)$ is not simply a constant multiple of the other, (3.24) is in fact the **general solution** of (3.23), and by choosing the constants A and B appropriately we may satisfy the *two* initial conditions which will typically accompany a second-order equation, namely

$$x = x_0, \qquad \frac{\mathrm{d}x}{\mathrm{d}t} = v_0 \quad \text{at} \quad t = 0, \tag{3.25}$$

x_0 and v_0 being given constants.

We stress that this powerful idea of linearly combining particular solutions to form a general solution *works only when the differential equation itself is linear.*

Example: constant coefficients

One of the most important special cases of (3.23) is

$$\frac{\mathrm{d}^2x}{\mathrm{d}t^2}+\beta x=0, \tag{3.26}$$

where β is a constant.

If $\beta > 0$ we may write $\omega = \beta^{1/2}$, so that

$$\frac{\mathrm{d}^2x}{\mathrm{d}t^2}+\omega^2 x=0. \tag{3.27}$$

One solution of this equation is $x_1 = \cos \omega t$, because $\dot{x}_1 = -\omega \sin \omega t$, so $\ddot{x}_1 = -\omega^2 \cos \omega t = -\omega^2 x_1$. Similarly, $x_2 = \sin \omega t$ is a solution, so

$$x = A \cos \omega t + B \sin \omega t \tag{3.28}$$

is the general solution.

If $\beta < 0$, on the other hand, we may write $q = (-\beta)^{1/2}$, so that

$$\frac{\mathrm{d}^2 x}{\mathrm{d}t^2} - q^2 x = 0. \tag{3.29}$$

Now $x_1 = \mathrm{e}^{qt}$ is one solution, because $\dot{x}_1 = q\,\mathrm{e}^{qt}$ and therefore $\ddot{x}_1 = q^2\,\mathrm{e}^{qt} = q^2 x_1$. Similarly, $x_2 = \mathrm{e}^{-qt}$ is a solution, so the general solution is now

$$x = C\,\mathrm{e}^{qt} + D\,\mathrm{e}^{-qt}, \tag{3.30}$$

where C and D are arbitrary constants.

Solutions to (3.26) are therefore of quite different types, depending on the sign of β; *oscillations* such as those in (3.28) are pursued further in Chapter 5, while (3.30), which shows $|x|$ typically growing without bound as $t \to \infty$, is central to the ideas of *instability* discussed in Chapter 10.

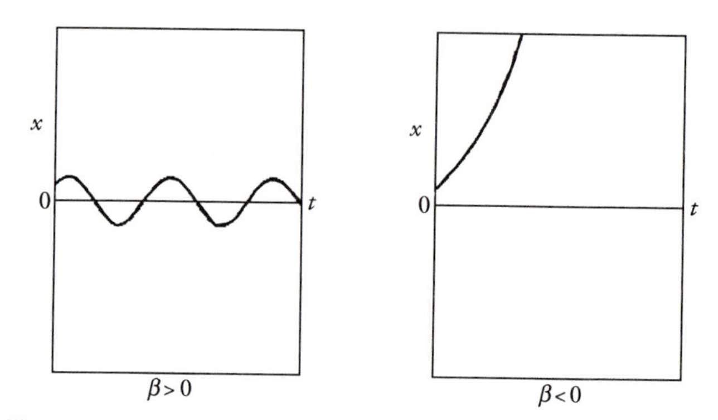

Fig. 3.7 Typical solutions to (3.26), depending on whether $\beta > 0$ or $\beta < 0$.

More generally, so long as a, b and c are *constants* in (3.23), $x = \mathrm{e}^{mt}$ is a solution if

$$am^2 + bm + c = 0. \tag{3.31}$$

Denoting the roots of this quadratic by m_1 and m_2 (which may be complex), we then have two particular solutions $\mathrm{e}^{m_1 t}$ and $\mathrm{e}^{m_2 t}$, so the general solution may be written

$$x = E\,\mathrm{e}^{m_1 t} + F\,\mathrm{e}^{m_2 t}, \tag{3.32}$$

unless it so happens that m_1 and m_2 are equal (Ex. 3.4).

Non-homogeneous linear equations

Suppose that we can find *one* solution, $x_{\mathrm{p}}(t)$ say, of the non-homogeneous linear equation (3.22), so that we have one particular function $x_p(t)$ such that

$$a\ddot{x}_p + b\dot{x}_p + cx_p = d. \tag{3.33}$$

Then on defining

$$u = x - x_p(t) \tag{3.34}$$

we find by subtracting (3.33) from (3.22) that

$$a\ddot{u} + b\dot{u} + cu = 0, \tag{3.35}$$

so that u satisfies the corresponding homogeneous equation (3.23).

We may therefore deal with non-homogeneous linear equations by finding the general solution u of the associated homogeneous problem and then adding a **particular integral** $x_p(t)$ (Ex. 3.5). This procedure works, again, only because the equation in question, (3.22), is linear.

3.5 Second-order nonlinear equations

We have noted already that a major distinction to be made with any non-linear equation is whether or not it is *autonomous*.

Autonomous equations

A nonlinear second-order equation of this type is of the form

$$\frac{\mathrm{d}^2x}{\mathrm{d}t^2} = f\left(x, \frac{\mathrm{d}x}{\mathrm{d}t}\right), \tag{3.36}$$

where f is some nonlinear function of the variables x and $\mathrm{d}x/\mathrm{d}t$, but with no explicit dependence on t. An example is the so-called van der Pol equation

$$\frac{\mathrm{d}^2x}{\mathrm{d}t^2} + \varepsilon(x^2-1)\frac{\mathrm{d}x}{\mathrm{d}t} + x = 0, \tag{3.37}$$

where the nonlinearity comes in through the term $x^2\,\mathrm{d}x/\mathrm{d}t$. This equation arises in connection with a certain type of electrical circuit (see Section 11.2).

It is often fruitful to recast an equation of the form (3.36) as a *pair* of coupled *first*-order equations, by writing

$$\frac{\mathrm{d}x}{\mathrm{d}t} = v, \qquad \frac{\mathrm{d}v}{\mathrm{d}t} = f(x, v). \tag{3.38a, b}$$

We may then use the chain rule (2.9) to eliminate t altogether from (3.38):

$$\frac{\mathrm{d}v}{\mathrm{d}x} = \frac{f(x, v)}{v}. \tag{3.39}$$

In this way we obtain a first-order differential equation for v as a function of x. It may be possible to solve this, giving $\mathrm{d}x/\mathrm{d}t$ $(=v)$ as a function of x. This then leaves us with a final first-order equation to be solved for x as a function of t.

Special case: $f(x, v)$ a function of x only

It may happen that (3.36) is of the still simpler form

$$\frac{\mathrm{d}^2x}{\mathrm{d}t^2}=f(x), \tag{3.40}$$

and in this case *some* progress with (3.39) can certainly be made, because that equation is then *separable*.

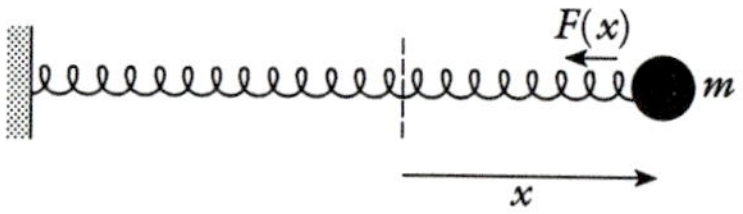

Fig. 3.8 A mass on a spring.

One example of this case arises when a mass m is attached to a spring which exerts a force $F(x)$ depending on the amount x by which it has been extended (Fig. 3.8). The equation of motion is then

$$m\frac{\mathrm{d}^2x}{\mathrm{d}t^2}=-F(x), \tag{3.41}$$

so (3.39) becomes

$$\frac{\mathrm{d}v}{\mathrm{d}x}=-\frac{F(x)}{mv}. \tag{3.42}$$

This is separable (see (3.18)), so

$$\int mv\,\mathrm{d}v=-\int F(x)\,\mathrm{d}x,$$

and therefore

$$\tfrac{1}{2}mv^2+\int F(x)\,\mathrm{d}x=\text{constant}. \tag{3.43}$$

If we now define the function

$$V(x)=\int_0^x F(s)\,\mathrm{d}s, \tag{3.44}$$

we may use (3.38a) to rewrite (3.43) in the form

$$\tfrac{1}{2}m\dot{x}^2+V(x)=\text{constant}. \tag{3.45}$$

We note in passing that this has a simple physical interpretation in terms of **conservation of energy**, with $\frac{1}{2}m\dot{x}^2$ denoting the kinetic energy of the mass and $V(x)$, defined by (3.44), denoting the potential energy of the spring.

If the initial conditions are, say,

$$x = x_0, \qquad \dot{x} = 0 \qquad \text{at} \quad t = 0, \tag{3.46}$$

then the constant in (3.45) must be $V(x_0)$, so

$$\frac{\mathrm{d}x}{\mathrm{d}t} = \pm\sqrt{\frac{2}{m}\{V(x_0) - V(x)\}}\,. \tag{3.47}$$

This is, again, separable, so

$$\pm\int_{x_0}^{x} \frac{\mathrm{d}x}{\sqrt{\dfrac{2}{m}\{V(x_0) - V(x)\}}} = t. \tag{3.48}$$

If the known function $V(x)$ is such that we can perform this final integration we will then have a direct relationship between x and t, as desired.

Non-autonomous equations

One example of such an equation is

$$\ddot{x} + k\dot{x} + x^3 = A\cos\Omega t, \tag{3.49}$$

where k, A and Ω are constants. This equation is of some practical interest, but it is nonlinear because of the term x^3 and non-autonomous as a result of the explicit dependence on t introduced by the term $A\cos\Omega t$.

Exact solutions to second-order equations of this kind are very rare, and this is not unrelated to the fact that the dependence of x on t in such cases may possibly be *chaotic* (see Section 11.1).

3.6 Phase space

Finally, suppose that we have some dynamical problem, and that we manage to represent it mathematically as a *system* of first-order differential equations of the following kind:

$$\begin{aligned}
\dot{x}_1 &= f_1(x_1, x_2, \ldots, x_N) \\
\dot{x}_2 &= f_2(x_1, x_2, \ldots, x_N) \\
&\vdots \\
\dot{x}_N &= f_N(x_1, x_2, \ldots, x_N).
\end{aligned} \tag{3.50}$$

Note that this system is *autonomous*; time t does not appear explicitly.

We then say that we are working in **phase space**, and the coordinates of this N-dimensional space are simply the variables $x_1, x_2, \ldots, x_N$.

Let us take some examples. Suppose we have a problem involving a simple oscillator, so that

$$\frac{\mathrm{d}^2 x}{\mathrm{d}t^2} + \omega^2 x = 0 \tag{3.51}$$

(see (3.27)). By introducing $y = \mathrm{d}x/\mathrm{d}t$ we may recast this equation as

$$\begin{aligned} \dot{x} &= y, \\ \dot{y} &= -\omega^2 x, \end{aligned} \tag{3.52}$$

which is of the form (3.50). The phase space for this problem is therefore two-dimensional, and our chosen coordinates in that space are x and y.

If we take (3.5) instead, i.e.

$$\frac{\mathrm{d}x}{\mathrm{d}t} = -2x + t, \tag{3.53}$$

we might at first think that the phase space is only one-dimensional, but it is not, because (3.53) is not autonomous, in view of the explicit appearance of t. We may, however, turn (3.53) into an autonomous first-order system by the apparently trivial step of supplementing it with the equation $\mathrm{d}t/\mathrm{d}t = 1$:

$$\begin{aligned} \dot{x} &= -2x + t, \\ \dot{t} &= 1. \end{aligned} \tag{3.54}$$

This system *is* autonomous, and of the form (3.50), because we have (somewhat deviously) elevated t in status to become one of the *dependent* variables, as well as being the independent variable of the problem. The phase space for (3.53) is therefore two-dimensional, and our chosen coordinates in that space are x and t.

In the same way, (3.49) can be recast in the form

$$\begin{aligned} \dot{x} &= y, \\ \dot{y} &= -ky - x^3 + A\cos\Omega t, \\ \dot{t} &= 1, \end{aligned} \tag{3.55}$$

and the associated phase space is three-dimensional.

The question that remains, obviously, is why we should go to the trouble of actually doing any of this. There are, in fact, at least three good reasons.

First, recasting an autonomous second-order equation as two first-order equations can sometimes help us to find an exact solution to the problem; we have already seen an example of this in Section 3.5.

Another, far deeper reason is that problems of the kind (3.50) lend themselves to essentially *geometric* arguments in phase space, and we shall see something of this in Section 5.5 and Chapter 11.

For present purposes, however, a major advantage of the system (3.50) is that it is in a most convenient form for a *computational* attack on the whole problem, and this is the subject of the next chapter.

Exercises

3.1 Solve

$$\frac{dx}{dt} + 2tx = t$$

subject to $x = 1$ when $t = 0$.

3.2 Solve

$$\frac{dx}{dt} = \frac{x^2}{1+t}$$

subject to $x = 1$ when $t = 0$. Does the solution 'blow up' in a finite time?

3.3 (a) Verify that (3.24) satisfies (3.23) for any values of the constants A and B.

(b) Solve

$$\ddot{x} - x = 0$$

subject to the initial conditions $x = 1, \dot{x} = 0$ when $t = 0$.

3.4 Consider the general homogeneous linear equation (3.23), i.e.

$$a\ddot{x} + b\dot{x} + cx = 0,$$

where a, b and c are given functions of t. Suppose that we have found one solution of this, $x = x_1(t)$, say, but have difficulty finding another. Show that by writing $x = x_1(t)u(t)$ we may reduce the equation to a first-order problem for the variable $z = \dot{u}$.

Use this method to solve

$$\ddot{x} - 2\dot{x} + x = 0$$

subject to $x = 1, \dot{x} = 0$ when $t = 0$, noting that the procedure leading to (3.31) gives only one solution, $x_1(t) = e^t$, in this particular case.

3.5 Solve

$$\ddot{x} - x = t$$

subject to $x = 1, \dot{x} = 0$ when $t = 0$.

3.6 *The simple harmonic oscillator*. In the case of a *linear* spring, with $F(x) = \alpha x$ in (3.41), α being the 'spring constant', we have

$$m \frac{d^2 x}{dt^2} = -\alpha x,$$

with, say, $x = x_0$ and $\dot{x} = 0$ when $t = 0$. Show that the (wholly elastic) potential energy is

$$V = \tfrac{1}{2}\alpha x^2,$$

and carry out the integration in (3.48) to obtain the solution. Check that this agrees with that obtained by using (3.26) and (3.28) instead.

4 Computer solution methods

4.1 Introduction

It is often quite impossible to solve a differential equation exactly, especially if the equation in question is nonlinear. Even if a 'solution' can be obtained, it may involve such awkward integrals or infinite series as to be virtually useless.

An example of this is provided by one of the earliest differential equations on record:

$$\frac{\mathrm{d}x}{\mathrm{d}t} = (1 + t)x + 1 - 3t + t^2 \tag{4.1}$$

(Newton 1671). This is first-order and linear, so we may tackle it, in principle, by the method of integrating factors (Section 3.2). In practice, however, we are unable to carry out the complicated final integration. We certainly get no hint at all from this approach of a major property of (4.1), namely that x

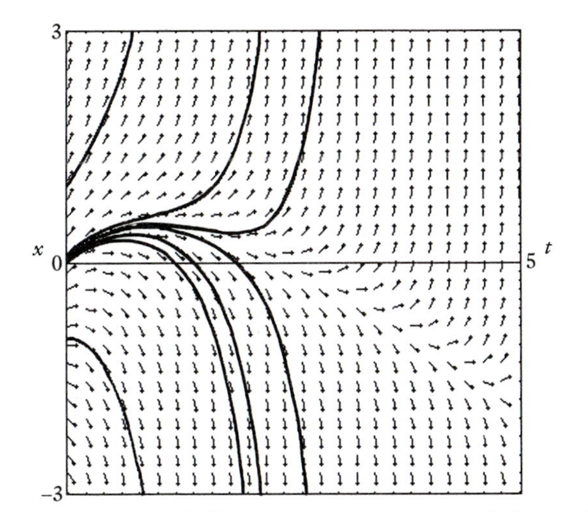

Fig. 4.1 Various solution curves to (4.1) obtained by a step-by-step method; $x_0 = -1, 0, 0.03, 0.06, 0.07, 0.10, 1$.

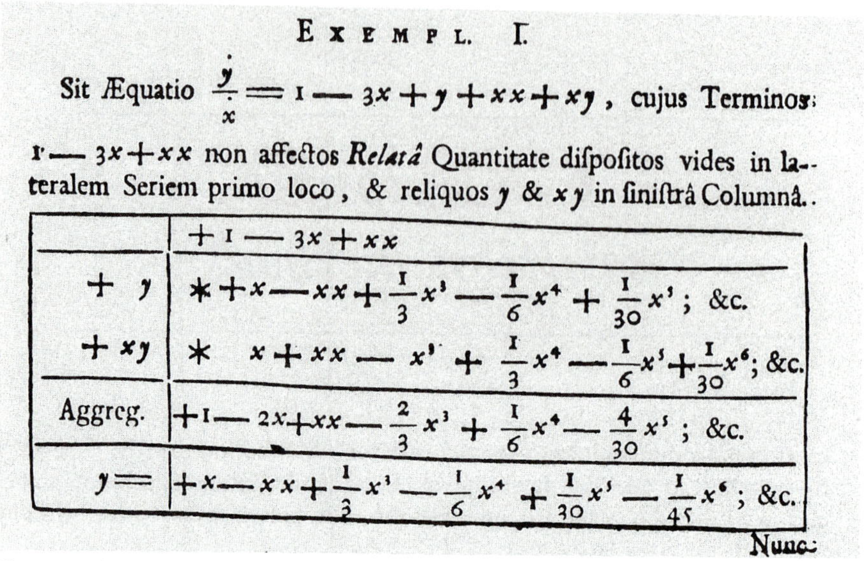

EXEMPL. I.

Sit Æquatio $\frac{\dot{y}}{\dot{x}} = 1 - 3x + y + xx + xy$, cujus Terminos $1 - 3x + xx$ non affectos *Relatâ* Quantitate dispositos vides in lateralem Seriem primo loco, & reliquos y & xy in sinistrâ Columnâ.

	$+1 - 3x + xx$
$+y$	$* + x - xx + \frac{1}{3}x^3 - \frac{1}{6}x^4 + \frac{1}{30}x^5$; &c.
$+xy$	$* \quad x + xx - x^3 + \frac{1}{3}x^4 - \frac{1}{6}x^5 + \frac{1}{30}x^6$; &c.
Aggreg.	$+1 - 2x + xx - \frac{2}{3}x^3 + \frac{1}{6}x^4 - \frac{4}{30}x^5$; &c.
$y =$	$+x - xx + \frac{1}{3}x^3 - \frac{1}{6}x^4 + \frac{1}{30}x^5 - \frac{1}{45}x^6$; &c.

Nunc:

Fig. 4.2 The differential equation (4.1), in Newton's Methodus Fluxionum et Serierum Infinitarum *of 1671. His infinite series approach gives, in the last line,* $x=t-t^2+\frac{1}{3}t^3-\frac{1}{6}t^4+\cdots$ *as the solution to (4.1) satisfying the initial condition* $x=0$ *when* $t=0$.

eventually tends to either $+\infty$ or $-\infty$ depending on whether its initial value x_0 is greater than or less than 0.066. This behaviour is, however, at once apparent from Fig. 4.1, which was obtained by using a *step-by-step* method on a computer.

We now consider three such methods, beginning with the simplest, which was devised by Euler in 1768. We apply each method, in the first instance, to the first-order problem

$$\frac{dx}{dt} = f(x,t), \quad \text{with } x = x_0 \text{ at } t = 0. \tag{4.2}$$

4.2 Euler's method

We know from the initial condition that $x = x_0$ when $t = 0$. Now, we also know from the equation (4.2) itself that the slope of the solution curve, dx/dt, is $f(x_0, 0)$ at $t = 0$. We may therefore obtain an approximation x_1 to the value of x at a small time h later by adding $hf(x_0, 0)$ to x_0, i.e.

$$x_1 = x_0 + hf(x_0, 0)$$

(Fig. 4.3). We may then take another step forward in the same way, using the slope $f(x,t)$ corresponding to the 'new' time $t_1 = h$ and our approximation x_1 to the value of x at that new time:

$$x_2 = x_1 + hf(x_1, t_1).$$

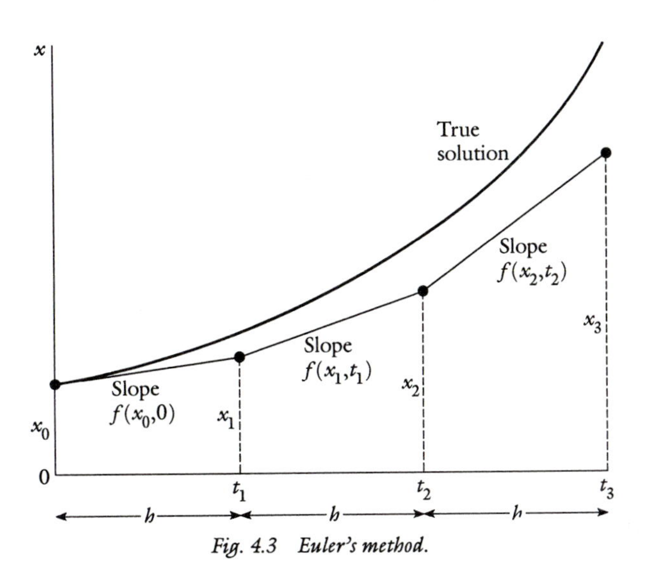

Fig. 4.3 Euler's method.

The idea of Euler's method, then, is to advance in time steps of amount h according to the rule

$$x_{n+1} = x_n + hf(x_n, t_n) \tag{4.3}$$

where x_0 is the given initial value of x and x_n is our approximation to the true value of x at time $t_n = nh$. To put it another way, we advance according to the rule

$$x_{\text{new}} = x_{\text{old}} + hf(x_{\text{old}}, t_{\text{old}}). \tag{4.4}$$

In Fig. 4.3 the procedure looks rather crude, because we have taken a large time step h in order that the procedure can be seen clearly. The essence of the method, however, is to obtain a good approximation to the solution at any given time t by taking a *large* number of *very small* steps, so that the constructed curve in Fig. 4.3 remains very close to the actual solution curve $x = x(t)$.

An example

There is one particular case in which we can give a simple, and most elegant, demonstration that Euler's method actually works. Specifically, we show that at *any fixed time* $t = nh$, the value x_n obtained by Euler's method tends to the correct value x *in the limit* $h \to 0$ (with, correspondingly, the number of steps $n \to \infty$, in order that $t = nh$ be fixed).

The particular case we have in mind is $f(x,t)=ax$, i.e.

$$\frac{\mathrm{d}x}{\mathrm{d}t}=ax, \qquad x=x_0 \text{ at } t=0, \tag{4.5}$$

where a is a constant. We know the exact solution to be

$$x=x_0\,\mathrm{e}^{at} \tag{4.6}$$

(see (3.4)).

Now, in this case $f(x_n,t_n)=ax_n$, so the recurrence relation (4.3) becomes

$$\begin{aligned} x_{n+1}&=x_n+hax_n \\ &=(1+ha)x_n. \end{aligned} \tag{4.7}$$

Most unusually, we may solve this at once to obtain a general expression for x_n in terms of n, for (4.7) states that each member of the sequence $x_0, x_1, x_2 \ldots$ is simply the previous member multiplied by a constant, $1+ha$. It follows, then, that

$$x_n=x_0(1+ha)^n, \tag{4.8}$$

and at time $t=nh$ we have

$$x_n=x_0\left(1+\frac{at}{n}\right)^n. \tag{4.9}$$

If we finally let $h\to 0$ *at fixed* t (with n tending to infinity, as $n=t/h$), we have

$$\lim_{\substack{h\to 0,\\ \text{fixed } t}} x_n = x_0 \lim_{\substack{n\to\infty,\\ \text{fixed } t}}\left(1+\frac{at}{n}\right)^n. \tag{4.10}$$

But, remarkably, the limit on the right-hand side is none other than (2.38), and we therefore know that

$$\lim_{\substack{h\to 0,\\ \text{fixed } t}} x_n = x_0\,\mathrm{e}^{at}, \tag{4.11}$$

in accord with (4.6). In this way we rigorously establish the validity of Euler's method in one particular case, when $f(x,t)=ax$ in (4.2).

The method can be shown to be valid for general $f(x,t)$, in the same sense (i.e. fixed t, $h\to 0$).

4.3 Computer implementation of Euler's method

Using Euler's method on a PC can be extremely straightforward, *even if the reader has no previous computing experience*. We will use the simple programming language QBasic, and Appendix A contains a start-from-scratch account of how to get programs working. Readers with little or no programming experience may find it helpful to work through Appendix A before proceeding further.

When ready, consider as an example the particular problem

$$\frac{dx}{dt} = x, \qquad x = 1 \text{ at } t = 0. \tag{4.12}$$

The short program

```
h = 0.01 : tm = 1
t = 0 : x = 1
 DO
   x = x + h * (x)
   t = t + h
 LOOP UNTIL ABS(t - tm) < h/2
PRINT t, x
```

is, in a sense, all that is required, and captures the essence of Euler's step-by-step method.

The first line specifies the time step `h` and the maximum value of `t` for which the step-by-step routine is to proceed, which we denote by `tm`, a sort of shorthand for 'tmax'. The second line sets the variables `t` and `x` equal to their initial values.

The heart of the program lies in the `DO...LOOP` which follows. The first line updates the value of `x` according to the rule (4.4), and if the right-hand side of (4.12) were some different function $f(x, t)$ we would of course enter the relevant expression between the brackets, in place of `x`. The next line updates `t` in the obvious way.

Most importantly, the `DO...LOOP` causes this updating of `x` and `t` to be *continually repeated* until at last the condition `ABS(t − tm) < h/2`, i.e. $|t - tm| < h/2$, is satisfied. The slightly curious form of this condition is in response to the fact that an integral number of chosen step lengths `h` will not in general fit in exactly to the chosen time interval `tm`, even before rounding errors in the computer have been taken into account. The given condition ensures that the computation ends at that discrete time $t = nh$ which is closest to the desired end-time `tm`, which is 1 in the above program. The last line causes the final values of `t` and `x` to appear on screen.

The exact solution to (4.12) is of course $x = e^t$, so the correct value of x at $t = 1$ is

$$e = 2.71828\ldots \tag{4.13}$$

By running the program on my own PC I obtained the results shown in Table 4.1. These demonstrate a general feature of Euler's method, namely that at fixed time t the error is essentially proportional to the step length h. In

Table 4.1 Some numerical results, by Euler's method, for the problem (4.12).

h	$x(1)$	Error
0.1	2.593743	−0.124539
0.01	2.704813	−0.013469
0.001	2.716920	−0.001361
0.0001	2.717872	−0.000410
0.00001	2.715541	−0.002741
0.000001	2.693228	−0.025053

particular, we see how reducing the step length by a factor of 10 at first reduces the error by a factor of 10.

This rule breaks down, however, if h is too small ($\lesssim 10^{-4}$ in the present example). This is not because of a failure in the method as such, but because rounding errors in the computer can always accumulate and lead to an inaccurate answer if the number of steps involved in any calculation becomes too large.

Displaying the solution graphically on screen

This is, again, quite easy to do; the following QBasic program has only four more lines than the one above, and contains no frills, but accomplishes the task satisfactorily:

```
CLS : SCREEN 9
xm = 200 : tm = 6
VIEW (250, 10) - (550, 300), 0, 9
WINDOW (0, 0) - (tm, xm)
 h = 0.01
 t = 0 : x = 1
  DO
   x = x + h * (x)
   t = t + h
   PSET (t, x)
  LOOP UNTIL ABS (t - tm) < h/2
```

The first line clears the screen and calls for high screen resolution. The `VIEW` command specifies which portion of the screen is to be used for graphical display, while the `WINDOW` command specifies within that region a coordinate system for our mathematics, by allocating coordinates `(0, 0)` to the bottom left point and `(tm, xm)` to the top right point. The `PSET` command puts a point on screen with coordinates `(t, x)`, and because this command is inside the `DO...LOOP` this is done at every time step.*

* Readers wishing to use some other version of BASIC should note that commands such as `CLS, SCREEN 9, VIEW, WINDOW` and `PSET` may need to be changed.

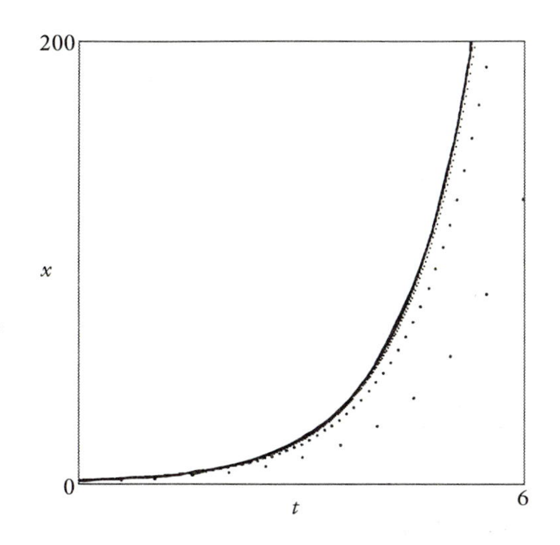

Fig. 4.4 Convergence to the solution of (4.12), by Euler's method. The lowest 'curve' was obtained with $h=0.5$, the one above with $h=0.1$. The curve with $h=0.02$ can be seen just beneath the one for $h=0.004$, which is indistinguishable from the exact solution on this scale.

Figure 4.4 shows the results of this program for four different step sizes h. As h is systematically decreased, the convergence to the true solution is quite evident.

A nonlinear example

The great merit of programs such as those above is that we may try them, with only minimal modification, on *any* problem of the same kind.

As our next example consider

$$\frac{dx}{dt} = t - x^2, \qquad x = x_0 \text{ when } t = 0, \tag{4.14}$$

which is nonlinear and non-autonomous, and has no exact solution in terms of elementary functions (despite its deceptively 'simple' appearance).

The key amendment to the above program comes, of course, in the line following the DO statement, which we replace by

```
x = x + h * (t - x ^ 2)
```
(4.15)

in keeping with (4.14). In order to plot both positive and negative values of x, where appropriate, we replace the WINDOW statement by the two lines

```
WINDOW (0, -xm) - (tm, xm)
LINE (0, 0) - (tm, 0), 9
```

the second of which draws the t-axis, in blue.

In Fig. 4.5 we display the results, with a step size of $h = 0.05$, for five

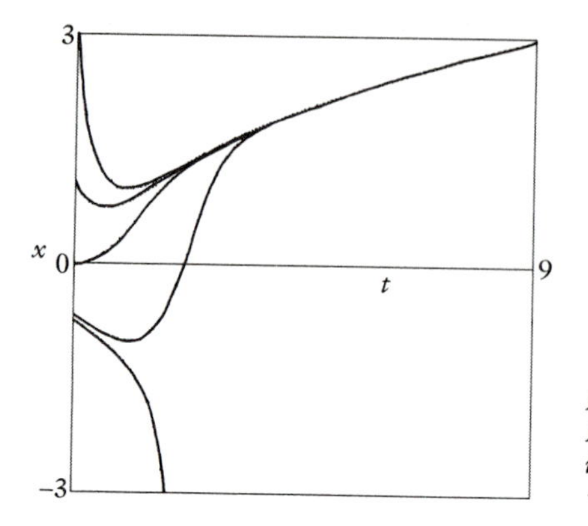

Fig. 4.5 Numerical solutions to (4.14), by Euler's method with h=0.05, for five different initial values x_0, namely 3, 1, 0, − 0.7 and − 0.75. Here tm=9 and xm=3.

different initial values x_0. If x_0 is less than about -0.73 the solution tends to $-\infty$, but for $x_0 \gtrsim -0.73$ the solutions corresponding to different initial conditions converge fairly rapidly to the same curve. If we run the integration a little longer, taking, say, `tm=36` and `xm=6` instead, we begin to suspect that this curve is $x = t^{1/2}$.

But if we run the program for too long we get something of a shock: beyond $t \sim 420$ large-amplitude oscillations set in, so that the computed value of x fluctuates dramatically at each time step. For $t \gtrsim 700$ these oscillations become chaotic (Fig. 4.6(a); see also Ex.11.6 and Fig. 11.17).

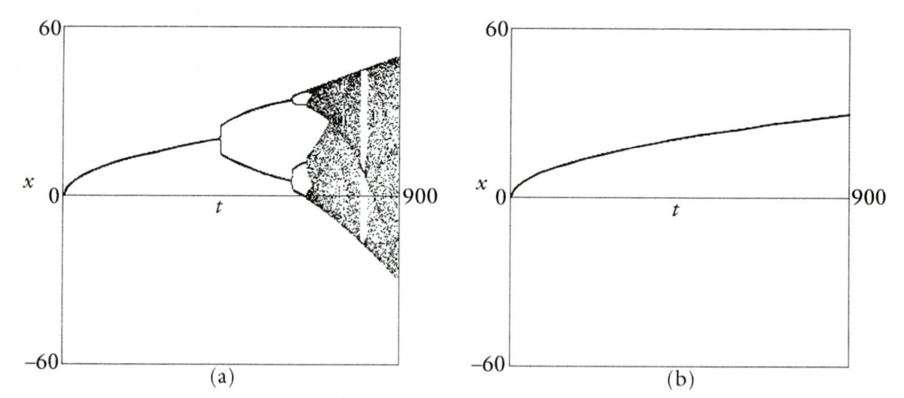

Fig. 4.6 Numerical integration of (4.14) over a longer time interval (a) with h=0.05, illustrating a dramatic breakdown of Euler's method, (b) with h=0.025. Here tm=900 and xm=60.

Now, this strange behaviour represents a breakdown of Euler's method, rather than a genuine property of the original differential equation (4.14), because it disappears if we repeat the computation with a smaller step size h (Fig. 4.6(b)).

Here, then, we have a sharp reminder that Euler's method, as explained in

Section 4.2, says nothing about fixing a fairly small h and then computing for a time $t = nh$ as large as we like. It involves, instead, fixing a *time* $t = nh$ and then, in principle, reworking the whole computation repeatedly, with smaller and smaller step sizes h and, consequently, more and more steps $n = t/h$.

In practice, we have to settle for rather less. *A simple rule of thumb with any step-by-step method is to keep halving the step size h until the results of successive computations, over a given time interval t, show no perceptible difference.*

We can do this quite simply by modifying our program a little further to:

```
CLS : SCREEN 9
xm = 60 : tm = 900
VIEW (250, 10) - (550, 300), 0, 9
WINDOW (0, -xm) - (tm, xm)
LINE (0, 0) - (tm, 0), 9
 DO
  INPUT "h, col"; h, col
  t = 0 : x = 0
   DO
    x = x + h * (t - x ^ 2)
    t = t + h
    PSET (t, x), col
   LOOP UNTIL ABS (t - tm) < h/2
 LOOP
```

As soon as the program is run, the `INPUT` command causes a request for the step size `h` and the colour number for `PSET` (an integer between 1 and 15, which we have called `col`) to appear on screen. If we type these in, separated by a comma, and press `RETURN`, the step-by-step method then proceeds, and when it has finished we are immediately prompted to try again, on the same screen with a different `h` and different colour, because we have embedded the main part of the program in a *second* `DO...LOOP`.

Our confidence in Fig. 4.6(b) comes, therefore, from the fact that using twice as many steps, with $h = 0.0125$, leads to a curve which is imperceptibly different, over the given time interval, from the one shown.

In the absence of any more sophisticated consideration of numerical errors in this book we will *always* apply this simple test, at least, to our computer 'results'.

4.4 Systems of differential equations

So far, we have applied Euler's method only to the problem

$$\dot{x} = f(x, t), \tag{4.16}$$

with $x = x_0$ at $t = 0$, but it extends in an obvious and entirely natural way to *systems* of first-order differential equations.

We might, for instance, have the coupled system

$$\begin{aligned} \dot{x} &= f(x, y, t), \\ \dot{y} &= g(x, y, t), \end{aligned} \tag{4.17}$$

f and g being given functions of x, y, and t. The problem would then be to find how *both* x and y evolve with time t, given, say, $x = x_0$ and $y = y_0$ at $t = 0$. In order to solve this numerically we have only to replace the rule (4.3) by its equivalent for the problem at hand:

$$\begin{aligned} x_{n+1} &= x_n + hf(x_n, y_n, t_n), \\ y_{n+1} &= y_n + hg(x_n, y_n, t_n). \end{aligned} \tag{4.18}$$

One common way in which a coupled system such as (4.17) arises is when we have a *second*-order equation

$$\ddot{x} = F(x, \dot{x}, t), \tag{4.19}$$

with $x = x_0$ and $\dot{x} = v_0$, say, at $t = 0$. This can always be recast as a coupled first-order problem:

$$\begin{aligned} \dot{x} &= y, \\ \dot{y} &= F(x, y, t), \end{aligned} \tag{4.20}$$

with $x = x_0$ and $y = v_0$ at $t = 0$ (cf. (3.38)). In this way the algorithm (4.18) allows us to solve second-order equations numerically.

A simple example

Consider as a test problem

$$\frac{\mathrm{d}^2 x}{\mathrm{d}t^2} + \omega^2 x = 0, \tag{4.21a}$$

with

$$x = 0, \qquad \frac{\mathrm{d}x}{\mathrm{d}t} = v_0 \qquad \text{at } t = 0, \tag{4.21b}$$

where ω and v_0 are given constants (cf. (3.27)). The exact solution is

$$x = \frac{v_0}{\omega} \sin \omega t, \tag{4.22}$$

but we shall solve the problem by Euler's method, for the sake of comparison.

Preliminary scaling of the variables

At first sight, perhaps, we are going to have to run the program again and again with various different numerical values for the constants ω and v_0. But this would in fact be a complete waste of time; by a simple transformation of the variables we may remove both ω and v_0 from the computational problem altogether.

To see this, define the new 'scaled' variables

$$\tilde{x} = \frac{x}{a}, \qquad \tilde{t} = \frac{t}{b}, \tag{4.23}$$

where a and b are *constants* which may be chosen at our convenience. Then

$$\frac{\mathrm{d}x}{\mathrm{d}t} = a\frac{\mathrm{d}\tilde{x}}{\mathrm{d}t} = a\frac{\mathrm{d}\tilde{x}}{\mathrm{d}\tilde{t}}\frac{\mathrm{d}\tilde{t}}{\mathrm{d}t} = \frac{a}{b}\frac{\mathrm{d}\tilde{x}}{\mathrm{d}\tilde{t}} \tag{4.24}$$

and so on, and in this way the problem (4.21) is transformed into

$$\frac{\mathrm{d}^2\tilde{x}}{\mathrm{d}\tilde{t}^2} + \omega^2 b^2 \tilde{x} = 0, \tag{4.25a}$$

with

$$\tilde{x} = 0, \qquad \frac{\mathrm{d}\tilde{x}}{\mathrm{d}\tilde{t}} = \frac{b}{a}v_0 \qquad \text{at } \tilde{t} = 0. \tag{4.25b}$$

Clearly, if we now choose $b = 1/\omega$ and $a = v_0/\omega$, so that

$$\tilde{x} = \frac{\omega}{v_0}x, \qquad \tilde{t} = \omega t, \tag{4.26}$$

the parameters ω and v_0 disappear from the computational problem altogether, for we are left with

$$\ddot{\tilde{x}} + \tilde{x} = 0, \tag{4.27a}$$

subject to

$$\tilde{x} = 0, \qquad \dot{\tilde{x}} = 1 \qquad \text{at} \quad \tilde{t} = 0. \tag{4.27b}$$

Here a dot denotes differentiation with respect to $\tilde{t}$.

Computational procedure

We first recast (4.27) as a first-order system:

$$\dot{\tilde{x}} = \tilde{y}, \tag{4.28a}$$

$$\dot{\tilde{y}} = -\tilde{x}, \tag{4.28b}$$

with $\tilde{x} = 0$, $\tilde{y} = 1$ when $\tilde{t} = 0$.

A simple QBasic program solving this problem by Euler's method, i.e. (4.18), is

```
CLS : SCREEN 9
xm = 2: tm = 5 * 2 * 3.14159
VIEW (250, 10) - (550, 300), 0, 9
WINDOW (0, -xm) - (tm, xm) : LINE (0, 0) - (tm, 0), 9
 h = .05
 t = 0: x = 0: y = 1
  DO
   dx = h * (y)
   dy = h * (-x)
   x = x + dx
   y = y + dy
   t = t + h
   PSET (t, x)
  LOOP UNTIL ABS(t - tm) < h/2
```

As written. this runs for five oscillation cycles of the exact solution to (4.27), which we know to be

$$\tilde{x} = \sin \tilde{t}. \tag{4.29}$$

This is equivalent to (4.22) by virtue of (4.26).

The results are shown in Fig. 4.7. With any given, i.e. fixed, time step $\tilde{h}$ the oscillations display a spurious increase in amplitude with time, which we know is not a property of the actual solution (4.29). This increase is, however, very gradual if $\tilde{h}$ is very small.

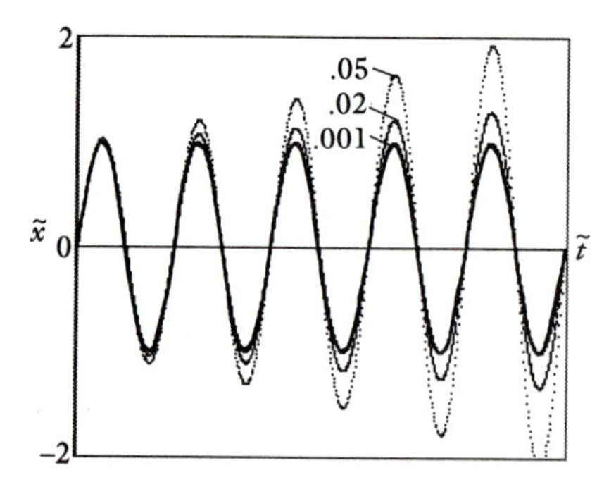

Fig. 4.7 Euler's method applied to (4.28), with $\tilde{x}=0$, $\tilde{y}=1$ at $\tilde{t}=0$, for three different (scaled) step sizes $\tilde{h}=\tilde{t}_{new}-\tilde{t}_{old}$.

More importantly, if we *fix the end-time*, the convergence of the computed solution to the actual solution as $\tilde{h}$ decreases is apparent.

4.5 More accurate step-by-step methods

Euler's method is conceptually simple, but the error at fixed time t is proportional to the time step h. If we require an accurate solution we therefore have to take h very small, and hence very many steps. This not only increases computation time, but risks the accumulation of rounding errors, as Table 4.1 shows.

We now consider two rather more sophisticated step-by-step methods. At fixed t the first has error proportional to h^2, and the second has error proportional to h^4, which, as h is small, represents a striking improvement on the Euler method.

Happily, both methods, like the Euler method itself, apply equally well to a single first-order equation or to a coupled system of such equations.

The improved Euler method

Let $x(t)$ denote the *exact* solution to the initial value problem

$$\frac{\mathrm{d}x}{\mathrm{d}t} = f(x,t) \qquad \text{with} \quad x = x_0 \quad \text{at} \quad t = 0. \tag{4.30}$$

Let $t_n = nh$ denote the discrete times at which we are seeking approximations x_n to the correct values $x(t_n)$.

Integrating both sides of (4.30) over one time step we find, without approximation, that

$$x(t_{n+1}) - x(t_n) = \int_{t_n}^{t_{n+1}} f[x(t),t]\,dt, \tag{4.31}$$

where $f[x(t),t]$ denotes the function of time only which is obtained when we substitute the exact solution $x(t)$ into the right-hand side of (4.30). If we draw the graph of $f[x(t),t]$, *rather than of* $x(t)$ *itself*, against t, we see that the right-hand side of (4.31) corresponds to the area under the curve, between t_n and t_{n+1} (Fig. 4.8).

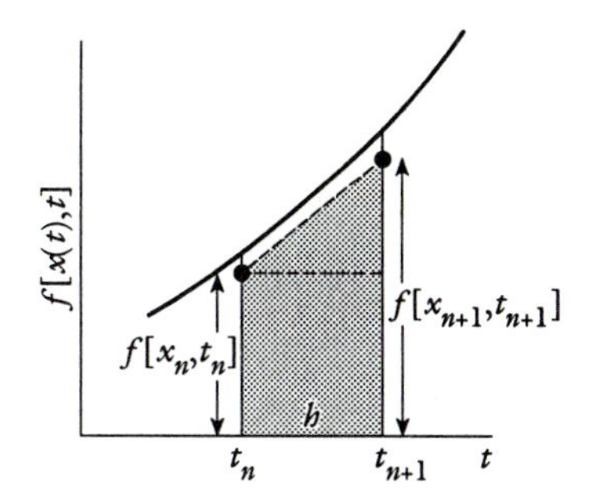

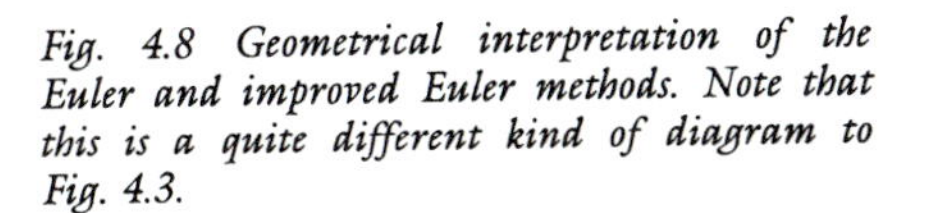
Fig. 4.8 Geometrical interpretation of the Euler and improved Euler methods. Note that this is a quite different kind of diagram to Fig. 4.3.

From this point of view, Euler's method consists in replacing the left-hand side of (4.31) by $x_{n+1} - x_n$, and replacing the right-hand side by an approximation to the area just mentioned, namely the area of the *rectangle* shown with height $f(x_n, t_n)$ and width h. In this way we may obtain

$$x_{n+1} = x_n + hf(x_n, t_n). \tag{4.32}$$

Clearly, however, a better approximation to the area in question may be obtained by using the *trapezium* in Fig. 4.8, which has area

$$\tfrac{1}{2}h\{f(x_n, t_n) + f(x_{n+1}, t_{n+1})\}. \tag{4.33}$$

This would give

$$x_{n+1} = x_n + \tfrac{1}{2}h\{f(x_n, t_n) + f(x_{n+1}, t_{n+1})\} \tag{4.34}$$

in place of (4.32), but this is not, as it stands, in a suitable form for computation, as it is not an explicit expression for the 'new' value x_{n+1} in terms of the 'old' value x_n.

To get over this difficulty we use the first approximation (4.32) to replace $f(x_{n+1}, t_{n+1})$ in (4.34) with $f\{x_n + hf(x_n, t_n), t_{n+1}\}$; (4.34) then becomes an explicit expression for x_{n+1} in terms of x_n.

In summary, there are three steps to the improved Euler method, and these are contrasted below with the Euler method itself:

Euler method	*Improved Euler method*	
$c_1 = hf(x,t)$	$c_1 = hf(x,t)$	(4.35)
	$c_2 = hf(x + c_1, t + h)$	
$x_{new} = x + c_1$	$x_{new} = x + \tfrac{1}{2}(c_1 + c_2)$.	

Note how, in the improved method, the computation of c_2 involves the just-computed quantity c_1.

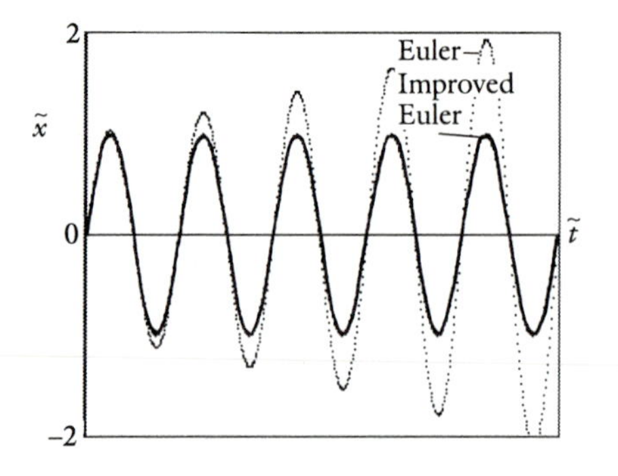

Fig. 4.9 Euler method v. improved Euler method: an example.

Figure 4.9 shows a comparison between the two methods, as applied to the initial value problem (4.27), both with a time step $h = 0.05$. The much better accuracy of the improved Euler method is clearly in evidence.

The Runge–Kutta method

This method is more accurate still, with an error proportional to h^4 at fixed time $t = nh$. It dates from the end of the last century and has a total of five steps in the up-dating process, each involving the quantity that has just been calculated in the step before:

$$\begin{aligned} c_1 &= hf(x, t) \\ c_2 &= hf(x + \tfrac{1}{2}c_1, t + \tfrac{1}{2}h) \\ c_3 &= hf(x + \tfrac{1}{2}c_2, t + \tfrac{1}{2}h) \\ c_4 &= hf(x + c_3, t + h) \\ x_{\text{new}} &= x + \tfrac{1}{6}(c_1 + 2c_2 + 2c_3 + c_4). \end{aligned} \tag{4.36}$$

We give some idea of how the Runge–Kutta method can be established in Ex. 4.6, but the actual derivation is beyond the scope of this book (and extremely lengthy). The best way forward, in the circumstances, is to check that the method works well on several differential equations *with known solutions*, as in Ex. 4.3 and Ex. 4.4, before we use it, later in the book, to venture further afield.

It is perhaps worth mentioning at this point that any of the three methods discussed above can be used, if desired, with a *variable step size h*. There is no need, in other words, for h to be constant throughout the whole computation from $t = 0$ to $t = \texttt{tm}$; we may use one value of h in (4.36) to update x_n to x_{n+1}, say, and then another, different value of h when we apply the algorithm again to update x_{n+1} to x_{n+2}. Adjusting the step size with time in this way can significantly reduce numerical errors, and an important example is given in Section 6.8.

We end this chapter by returning to the beginning, i.e. the differential equation (4.1), which Newton first considered in 1671. We try all three step-by-step methods on this problem, and make it quite deliberately taxing by choosing the initial condition $x_0 = 0.0655$ very close to the critical value of 0.065923 which separates those solutions which eventually tend to $+\infty$ from those which eventually tend to $-\infty$ (see Fig. 4.1).

With a common step size of $h = 0.035$ the Euler curve is satisfactory until $t \doteqdot 1$, but then takes a 'wrong turn', while the curve obtained by the improved Euler method turns the correct way but is substantially in error by $t \sim 3$. The curve obtained by the Runge–Kutta method, however, is indistinguishable from (what further computations indicate to be) the true solution, and it would be so, in fact, on the time scale indicated, even with a step size h as

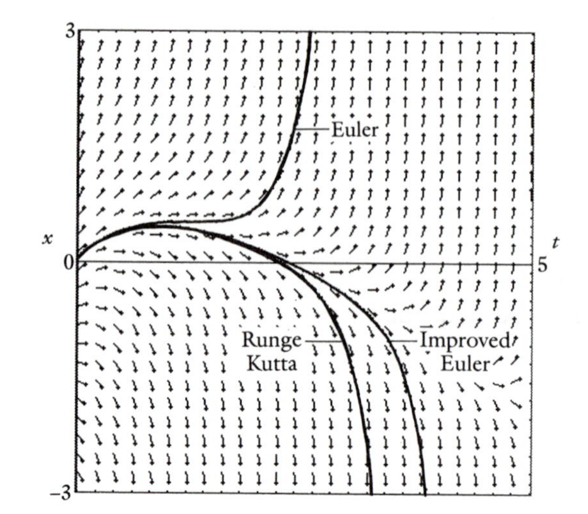

Fig. 4.10 Comparison of three step-by-step methods on equation (4.1), with $x_0 = 0.0655$ and $h = 0.035$ in each case.

large as 0.1. The additional accuracy of the Runge–Kutta method makes it the obvious choice for the more lengthy or delicate numerical integrations that are required later in this book.

Exercises

4.1 Type in, run and save the various Euler method programs in Section 4.3.

Modify the second program so that the exact solution $x = e^t$ is plotted on screen, in a different colour, for comparison with the Euler method approximation.

Does the smaller value of h used in Fig. 4.6(b) 'cure' the breakdown of Euler's method in Fig. 4.6(a), or simply postpone it till a later time t?

4.2 Type in, run and save the Euler method program in Section 4.4, and confirm the results in Fig. 4.7.

One might think, perhaps, that the updating process could be programmed more succinctly as

```
x = x + h * (y)
y = y + h * (-x),
```

but this would in fact be different from Euler's method. Why?

4.3 Consider again the problem

$$\frac{dx}{dt} = x, \qquad x = 1 \text{ when } t = 0.$$

Change the first program in Section 4.3 so that it uses the improved Euler method instead, and construct the equivalent of Table 4.1 for $h = 0.1$, 0.01, 0.001.

Then repeat the exercise using the Runge–Kutta method and computation with double precision accuracy (Section A.3).

$$[x(1) = \mathrm{e} = 2.71828\,18284\,59045\ldots]$$

4.4 Show by means of a little algebra that just *one* step (length h) of the Runge–Kutta method in the above problem leads to

$$x = 1 + h + \frac{h^2}{2!} + \frac{h^3}{3!} + \frac{h^4}{4!}$$

when $t = h$, which is just the Taylor approximation for e^h, correct to $O(h^4)$.

4.5 Confirm the results for the differential equation (4.1) in Figs 4.1 and 4.10 by using the program `1XT` in Appendix B, which can employ any of the three step-by-step methods and plot the direction field.

If we were using only the Euler method, how would we know that the corresponding 'result' in Fig. 4.10 is not reliable?

What evidence do we have that the Runge–Kutta curve in Fig. 4.10 *does* give the correct solution?

4.6 Let $x(t)$ satisfy the autonomous differential equation

$$\frac{\mathrm{d}x}{\mathrm{d}t} = f(x),$$

with $x = x_0$ at $t = 0$. Expand $x(t)$ in a Taylor series (see (2.14)) to show that if h is small

$$x(h) \doteqdot x_0 + hf(x_0) + \tfrac{1}{2}h^2 f(x_0)f'(x_0),$$

correct to order h^2. Then show that one step of the *improved* Euler method (4.35) gives the same result, correct to order h^2.

[Exactly the same approach, when carried out correct to *fourth* order, can be used to establish the validity of the Runge–Kutta method, but the algebra is daunting, even in the autonomous case.]

5 Elementary oscillations

5.1 Introduction

One of the oldest and best-known oscillating systems is the so-called **simple pendulum**. This consists of a light rigid rod of length l with a point mass at one end, the other end being pivoted at a fixed point O, so that the pendulum can swing freely in one particular vertical plane (Fig. 5.1).

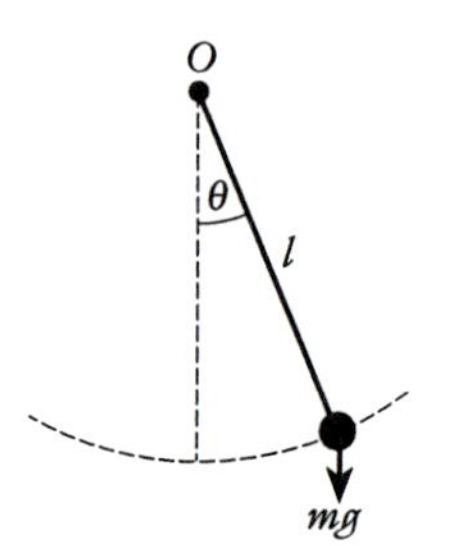

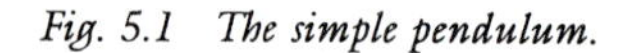

Fig. 5.1 The simple pendulum.

Galileo made careful observations of pendulums in about 1602, and found the period of oscillation T to be proportional to the square root of the length:

$$T \propto \sqrt{l}\,. \tag{5.1}$$

What really impressed him, however, was the way in which the oscillation period seemed to be independent of the amplitude, although it emerged subsequently that this is true only for *small* oscillations of the pendulum.

In order to set up the appropriate differential equation of motion, let θ (in radians) denote the angle between the pendulum and the downward vertical (Fig. 5.1). The bob has at any instant a velocity $l\,\mathrm{d}\theta/\mathrm{d}t$ in the direction of increasing θ, and its acceleration component in that direction is $l\,\mathrm{d}^2\theta/\mathrm{d}t^2$. The only force on the bob in that direction is a component $mg\sin\theta$ due to gravity, and this acts in the opposite sense, i.e. the direction of decreasing θ. So $ml\,\mathrm{d}^2\theta/\mathrm{d}t^2$ must be equal to $-mg\sin\theta$, i.e.

$$\frac{\mathrm{d}^2\theta}{\mathrm{d}t^2} + \frac{g}{l}\sin\theta = 0. \tag{5.2}$$

This is not easy to solve analytically, but if we confine attention to *small* swings of the pendulum we may use the approximation

$$\sin\theta \doteqdot \theta \qquad \text{for} \quad |\theta| \ll 1, \tag{5.3}$$

and we then have

$$\frac{d^2\theta}{dt^2} + \frac{g}{l}\theta = 0 \tag{5.4}$$

as the approximate equation of motion governing small-amplitude oscillations of a simple pendulum.

Suppose, then, that we draw the pendulum aside a small amount, to $\theta = \theta_0$, say, at $t = 0$, and release it from rest. The solution of (5.4) which satisfies these initial conditions is

$$\theta = \theta_0 \cos\left(\frac{g}{l}\right)^{1/2} t \tag{5.5}$$

(see (3.28)), so the pendulum swings to and fro with a period

$$T = 2\pi\sqrt{\frac{l}{g}}\,. \tag{5.6}$$

This is proportional to $\sqrt{l}$ and independent of the amplitude θ_0, as Galileo observed in his experiments.

5.2 The linear oscillator

The kind of simplification we saw in moving from (5.2) to (5.4)—called **linearization** of the original equation of motion—is typical of virtually any dynamical system when we restrict attention to *small displacements about a point of (stable) equilibrium.*

To see why this should be so, consider the point mass m in Fig. 5.2, which moves to and fro under the action of a spring. Let $x = 0$ denote the equilibrium point, at which the spring is neither extended nor compressed, and let the force exerted by the spring be $F(x)$ in the negative x-direction. In

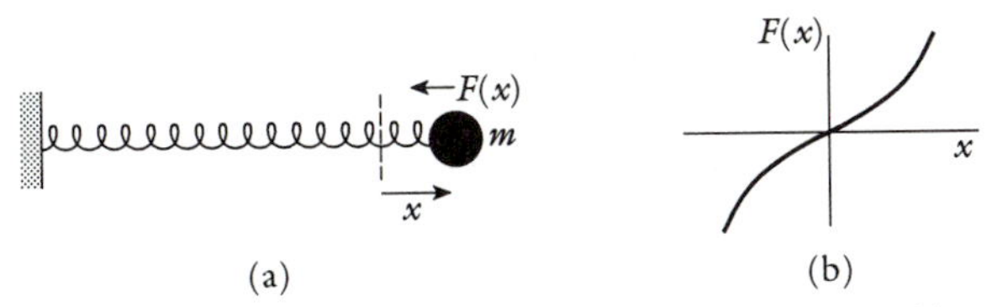

Fig. 5.2 (a) A spring oscillator. (b) A typical graph of spring force F(x) as a function of displacement x from equilibrium.

general this will be a complicated function of x, determined by the detailed elastic properties of the spring, but we do know that $F(0)=0$, because the spring force must be zero at the equilibrium point $x=0$.

Now, the exact equation of motion is

$$m\frac{d^2x}{dt^2} = -F(x), \tag{5.7}$$

but if we confine attention to *small* values of $|x|$, so that the particle is close to the equilibrium point, we may approximate $F(x)$ by the first two terms of its Taylor series about $x=0$:

$$F(x) \doteqdot F(0) + xF'(0) \tag{5.8}$$

(see (2.14)), and because $F(0)=0$ we may write

$$F(x) \doteqdot \alpha x, \tag{5.9}$$

where $\alpha = F'(0)$ denotes the positive 'spring constant.'

In this way, then, we again obtain a *linearized* equation of motion:

$$m\frac{d^2x}{dt^2} = -\alpha x, \tag{5.10}$$

approximately valid if $|x|$ is small. On writing

$$\omega^2 = \frac{\alpha}{m} \tag{5.11}$$

this becomes

$$\frac{d^2x}{dt^2} + \omega^2 x = 0, \tag{5.12}$$

with general solution

$$x = A\cos\omega t + B\sin\omega t \tag{5.13a}$$

(see (3.28)). This can also be written in the form

$$x = C\cos(\omega t - D), \tag{5.13b}$$

where $C=(A^2+B^2)^{1/2}$ and $D=\tan^{-1}(B/A)$.

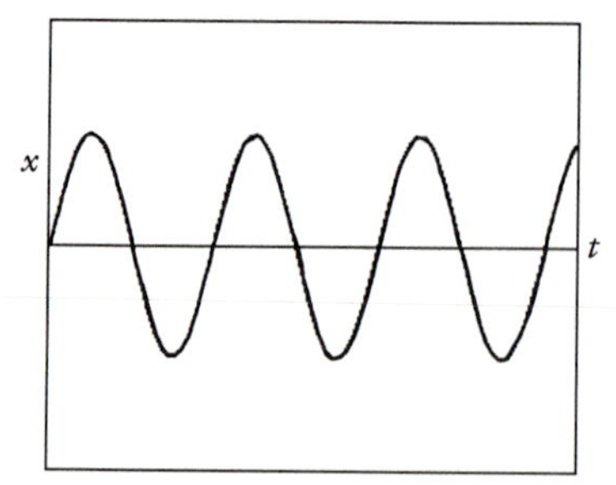

Fig. 5.3 A simple-harmonic oscillation.

Small oscillations about the equilibrium point are therefore simple harmonic (Fig. 5.3). The **period** of the oscillation is clearly $2\pi/\omega$, and the **frequency**, i.e. the number of oscillation cycles per unit time, is therefore $f=\omega/2\pi$. Having said this, we shall occasionally lapse into using the term 'frequency' to refer to ω itself; its proper title is *angular frequency*.

The effect of damping

Suppose now that the mass in Fig. 5.2 experiences also a frictional resistance which is proportional to its speed. This implies an additional force of $-\gamma\dot{x}$ in the positive x-direction, γ being a positive constant, and so

$$m\ddot{x} = -\alpha x - \gamma\dot{x}$$

is the new (approximate) equation of motion for small displacements x. On defining $k=\gamma/m$ we may rewrite this as

$$\ddot{x} + k\dot{x} + \omega^2 x = 0, \qquad (5.14)$$

where ω is defined, as before, by (5.11).

It may be shown that the general solution to this equation is

$$x = Ce^{-kt/2}\cos\left\{\left(\omega^2 - \tfrac{1}{4}k^2\right)^{1/2} t - D\right\}, \qquad (5.15)$$

provided that the damping is not too large, i.e. provided that $\frac{1}{4}k^2 < \omega^2$ (Ex. 5.1). The amplitude of the oscillations now decreases with time, in proportion to $e^{-kt/2}$, and the larger the frictional constant k the faster the decay of the oscillations (Fig. 5.4). A secondary effect of the damping is to reduce the oscillation frequency from ω to $(\omega^2 - \frac{1}{4}k^2)^{1/2}$.

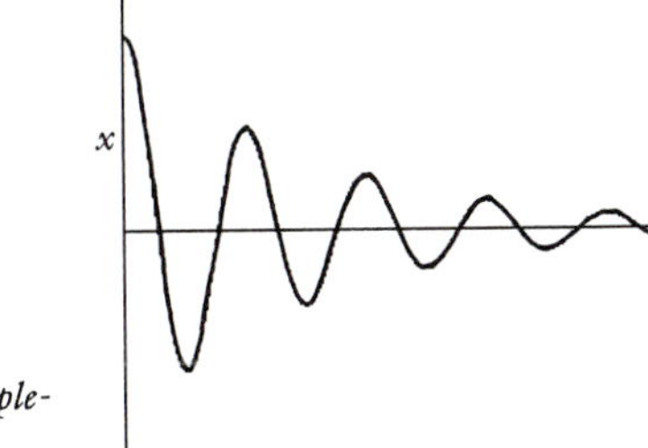

Fig. 5.4 An example of damped simple-harmonic motion.

Forced oscillations; resonance

Suppose finally that the particle in Fig. 5.2 is subject not only to the spring force $-\alpha x$ but to a prescribed *driving* force $F_0 \cos \Omega t$, which is itself oscillatory. In the absence of damping the equation of motion for small $|x|$ is then

$$\ddot{x} + \omega^2 x = a \cos \Omega t, \qquad (5.16)$$

where $a = F_0/m$.

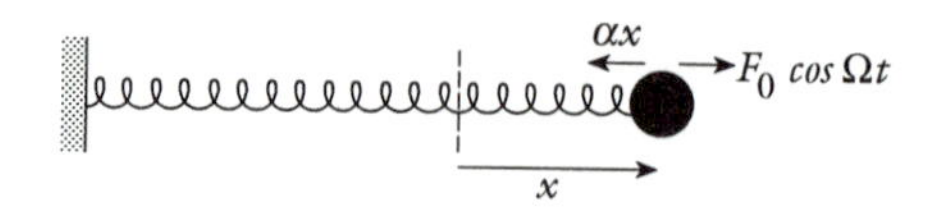

Fig. 5.5 The forced linear spring.

This equation was studied by Euler in 1739, and he noted that the most interesting case is when the forcing frequency Ω coincides with the natural frequency of the spring ω. In this case we have the phenomenon of **resonance**, resulting in steadily growing oscillations. When $\Omega = \omega$ we may verify, for example, that

$$x = \frac{a}{2\omega} t \sin \omega t \tag{5.17}$$

satisfies (5.16) and the initial conditions $x = \dot{x} = 0$ at $t = 0$ (Ex. 5.2), so that the amplitude of the oscillation increases in proportion to t as time goes on.

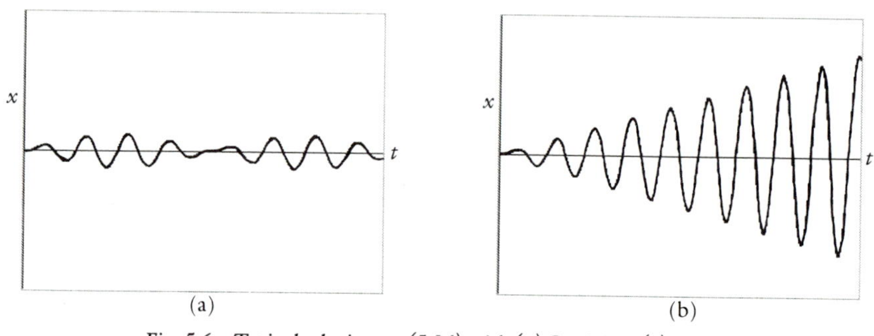

Fig. 5.6 Typical solutions to (5.16) with (a) $\Omega = 0.8\omega$, (b) $\Omega = \omega$.

If a little damping is present, the amplitude begins by growing in this way, but eventually settles down to a (relatively large) constant value (Fig. 5.7(a)). This final steady state response is largest when Ω is close to the resonant value ω (Fig. 5.7(b)).

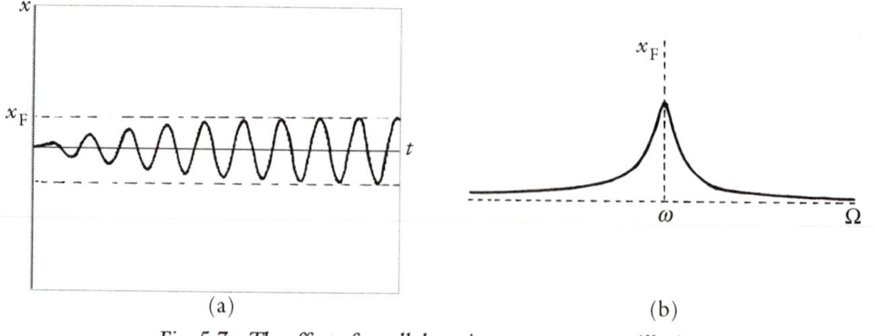

Fig. 5.7 The effect of small damping on resonant oscillations.

In practice, resonant oscillations may well be large enough that the linearized equation (5.16) ceases to be valid, because the approximation (5.9) breaks down.

5.3 Multiple modes of oscillation

Suppose now that two particles of mass m are attached to three identical springs of natural length a and constrained to move along a straight line, as in Fig. 5.8. We now need *two* coordinates to describe the system at any time t, and the system is said to have two **degrees of freedom**. If x_1 and x_2 denote small displacements from equilibrium, and if each spring has 'spring constant' α, then the tensions in the springs are

$$T_1 = \alpha x_1, \qquad T_2 = \alpha(x_2 - x_1), \qquad T_3 = -\alpha x_2, \tag{5.18}$$

because x_1, $x_2 - x_1$ and $-x_2$ are the amounts by which each spring is extended.

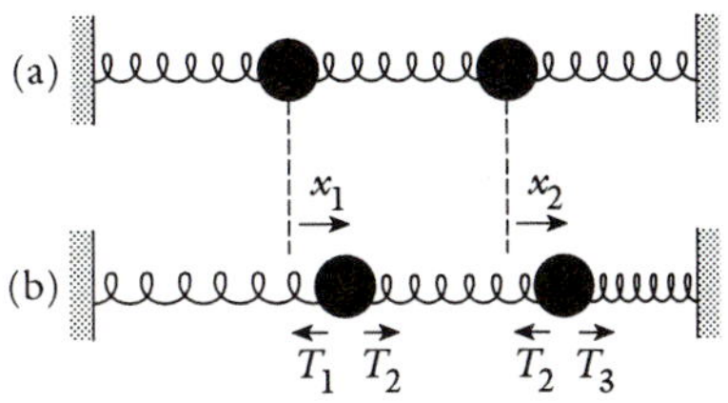

Fig. 5.8 An oscillating system with two degrees of freedom.

At time t, therefore, the first mass experiences a force $T_2 - T_1 = \alpha(x_2 - 2x_1)$, while the second experiences a force $T_3 - T_2 = \alpha(x_1 - 2x_2)$. The equations of motion are then

$$\begin{aligned} m\ddot{x}_1 &= \alpha(x_2 - 2x_1), \\ m\ddot{x}_2 &= \alpha(x_1 - 2x_2), \end{aligned} \tag{5.19a,b}$$

which are simultaneous differential equations for the two unknowns $x_1(t)$ and $x_2(t)$.

An obvious question which comes to mind is whether this system has a natural frequency of oscillation, as in the case of one degree of freedom (see (5.11) and Fig. 5.2(a)). We try, therefore,

$$x_1 = A\cos\omega t, \qquad x_2 = B\cos\omega t \tag{5.20}$$

and find that this does indeed represent a solution of (5.19a,b) if the constants A, B and ω are such that

$$\begin{aligned} -m\omega^2 A &= \alpha(B - 2A), \\ -m\omega^2 B &= \alpha(A - 2B). \end{aligned}$$

After a little rearrangement these equations become

$$\begin{aligned}\left(2-\frac{m}{\alpha}\,\omega^2\right)A&=B,\\ A&=\left(2-\frac{m}{\alpha}\,\omega^2\right)B,\end{aligned} \tag{5.21}$$

and unless $A=B=0$ we deduce that

$$\left(2-\frac{m}{\alpha}\,\omega^2\right)^2=1. \tag{5.22}$$

Thus $2-m\omega^2/\alpha=\pm 1$, i.e.

$$\omega^2=\frac{\alpha}{m} \quad \text{or} \quad \omega^2=\frac{3\alpha}{m}. \tag{5.23a, b}$$

On returning to (5.21) we see that in the first case $A=B$, but in the second case $A=-B$.

We therefore find that the system in Fig. 5.8 has *two* natural modes of oscillation: a 'low'-frequency mode in which both masses are travelling in the same direction at any particular time t, and a 'high'-frequency mode in which they are always travelling in opposite directions (Fig. 5.9).

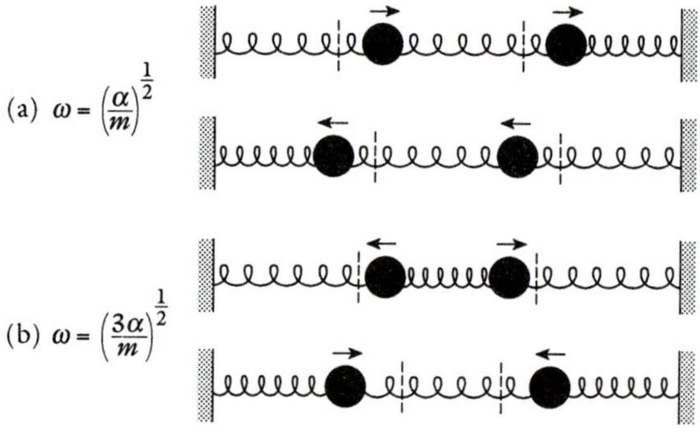

(a) $\omega=\left(\frac{\alpha}{m}\right)^{\frac{1}{2}}$

(b) $\omega=\left(\frac{3\alpha}{m}\right)^{\frac{1}{2}}$

Fig. 5.9 The two natural modes of oscillation for the system in Fig. 5.8.

We note, however that it is only for rather special initial conditions that the system will behave either as in Fig. 5.9(a) or as in Fig. 5.9(b); in general the response of the system will involve a linear combination of both oscillations, and will consequently appear to be rather more complicated.

The double and triple pendulum

An even more classical problem involving two natural modes of oscillation is the so-called **double pendulum**, in which one pendulum is suspended from the bob of another, both being constrained to swing in the same vertical plane. This problem is important if only because Euler and Daniel Bernoulli

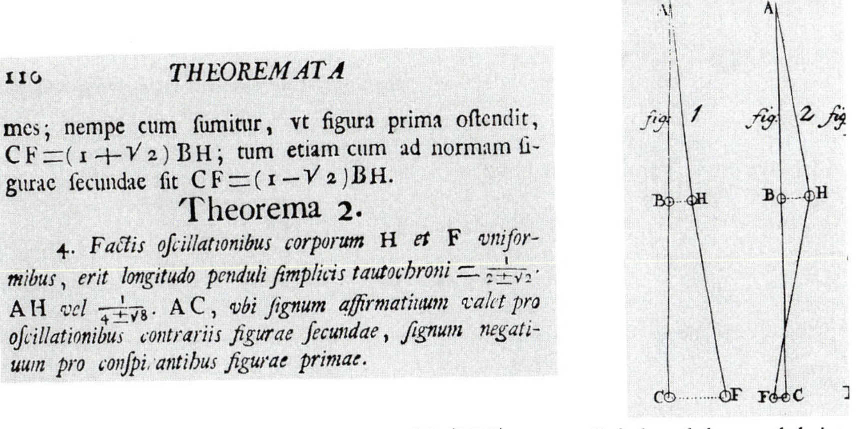

110 *THEOREMATA*

mes; nempe cum ſumitur, vt figura prima oſtendit, $CF=(1+\sqrt{2})BH$; tum etiam cum ad normam figurae ſecundae fit $CF=(1-\sqrt{2})BH$.

Theorema 2.

4. *Factis oſcillationibus corporum* H *et* F *vniformibus, erit longitudo penduli ſimplicis tautochroni* $= \frac{1}{2\pm\sqrt{2}}$ AH *vel* $\frac{1}{4\pm\sqrt{8}}$ · AC, *vbi ſignum affirmatiuum valet pro oſcillationibus contrariis figurae ſecundae, ſignum negatiuum pro conſpirantibus figurae primae.*

Fig. 5.10 Extract and figure from D. Bernoulli's (1738) paper on linked pendulums and chains.

happened to be working on it when, in the mid-1730s, it finally became clear that differential equations of motion would be the key to virtually all dynamical problems.

In any event, they found a 'long'-period oscillation in which the two pendulums swing to and fro in phase, and a 'short'-period oscillation in which they swing to and fro exactly out of phase, so that when one is swinging to the right the other is swinging to the left. When the two pendulums are of equal length l, with bobs of equal mass, the two periods are given by

$$T = 2\pi\left(1 \pm \frac{1}{\sqrt{2}}\right)^{1/2} \sqrt{\left(\frac{l}{g}\right)}, \tag{5.24}$$

and in the corresponding motions

$$\theta_2 = \pm\sqrt{2}\,\theta_1, \tag{5.25}$$

where θ_1 and θ_2 denote the angles made with the downward vertical by the upper and lower pendulums respectively (Fig. 5.10).

In a similar way the *triple* pendulum has *three* natural modes of oscillation, each with its own distinct frequency, and these are shown in Fig. 12.7, which is taken, again, from Daniel Bernoulli's original paper of 1738.

5.4 Coupled oscillators

When two identical oscillators are weakly coupled there can be a slow but striking transfer of energy from one to the other, and then back again.

For an example of this we may simply modify the system in Fig. 5.8 by making the middle spring much weaker than the other two, with spring constant $\varepsilon\alpha$, say, where $\varepsilon \ll 1$.

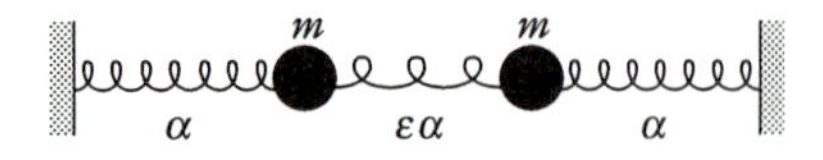

Fig. 5.11 Two weakly-coupled spring oscillators.

In place of (5.19) we then have

$$m\ddot{x}_1 = \alpha[\varepsilon x_2 - (1+\varepsilon)x_1],$$
$$m\ddot{x}_2 = \alpha[\varepsilon x_1 - (1+\varepsilon)x_2] \qquad (5.26\text{a, b})$$

as the equations of motion, and

$$x_1 = A\cos\omega t, \qquad x_2 = B\cos\omega t \qquad (5.27)$$

represents a possible solution if ω takes one of two values, ω_S or ω_F, where

$$\omega_S = \left(\frac{\alpha}{m}\right)^{1/2}, \qquad \omega_F = \left(\frac{\alpha}{m}\right)^{1/2}(1+2\varepsilon)^{1/2}. \qquad (5.28\text{a, b})$$

For the 'slow' mode, ω_S, we find $A = B$, while for the 'fast' mode, ω_F, we find $A = -B$. In the special case $\varepsilon = 1$ these results reduce to those in Section 5.3, as they should.

We are now interested in small values of ε, however, and in this case we may expand (5.28b) binomially—or in a Taylor series about $\varepsilon = 0$—to obtain

$$\omega_S = \left(\frac{\alpha}{m}\right)^{1/2}, \qquad \omega_F \doteqdot \left(\frac{\alpha}{m}\right)^{1/2}(1+\varepsilon) \qquad (5.29\text{a, b})$$

as a good first approximation, so that the 'fast' mode has only a slightly higher frequency than the 'slow' mode.

Consider next an initial value problem, in which one mass is slightly displaced at $t = 0$, but the other is not, both being originally at rest. Suppose, in other words, that

$$\left.\begin{array}{ll} x_1 = x_0, & \dot{x}_1 = 0 \\ x_2 = 0, & \dot{x}_2 = 0 \end{array}\right\} \text{ at } t = 0. \qquad (5.30)$$

We may verify that

$$x_1 = \tfrac{1}{2}x_0(\cos\omega_S t + \cos\omega_F t)$$
$$x_2 = \tfrac{1}{2}x_0(\cos\omega_S t - \cos\omega_F t), \qquad (5.31)$$

satisfies the equations of motion (5.26) and the initial conditions (5.30). The solution consists, therefore, of a linear superposition of both modes of oscillation.

The nature of the solution (5.31) becomes clearer if we use some elementary trigonometric identities together with the approximations $\omega_F + \omega_S \doteqdot 2\,\omega_S$, $\omega_F - \omega_S \doteqdot \varepsilon\omega_S$ to rewrite it as

$$\begin{aligned} x_1 &\doteqdot x_0 \cos \omega_S t \cos \tfrac{1}{2}\varepsilon\omega_S t, \\ x_2 &\doteqdot x_0 \sin \omega_S t \sin \tfrac{1}{2}\varepsilon\omega_S t. \end{aligned} \qquad (5.32\text{a, b})$$

Thus x_1 is essentially a simple harmonic oscillation with frequency ω_S but with an amplitude $x_0 \cos \tfrac{1}{2}\varepsilon\omega_S t$ which itself varies in a simple harmonic manner, though much more gradually, on account of the smallness of ε. In particular, the amplitude decays slowly to zero at a time $t = \pi/\varepsilon\omega_S$, and by this time the oscillations of x_2 have slowly built up to a maximum (see (5.32b) and Fig. 5.12). As time continues the pattern reverses, so that one complete to-and-fro transfer of energy takes place in a time $2\pi/\varepsilon\omega_S$, during which each individual oscillator will have executed $1/\varepsilon$ cycles of period $2\pi/\omega_S$.

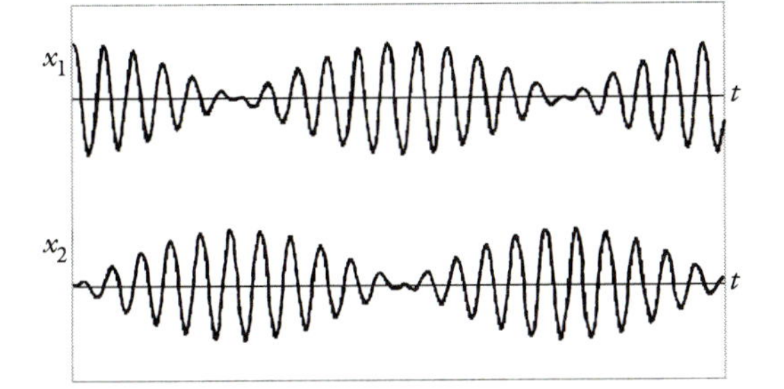

Fig. 5.12 Exchange of energy between the weakly-coupled spring oscillators in Fig. 5.11; $\varepsilon = 0.1$.

Two weakly coupled pendulums

For a simple practical demonstration of the above phenomenon, take a length of string and attach its ends to two fixed points P, S of equal height. Then suspend two string pendulums of equal length l, with identical bobs, from symmetrical points Q, R of the string, as in Fig. 5.13.

Now set one pendulum swinging, in a plane perpendicular to the plane of

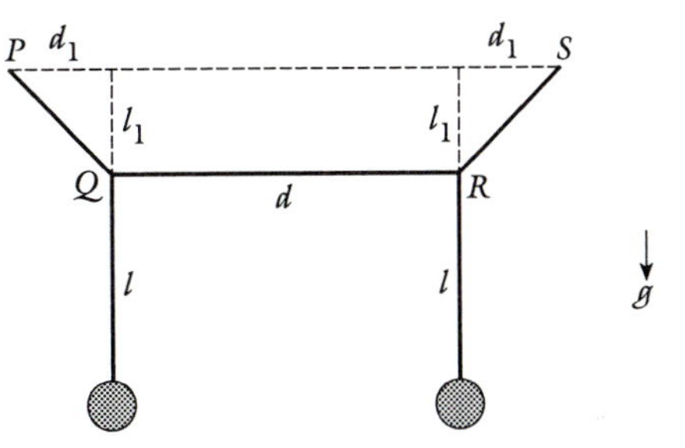

Fig. 5.13 Two weakly-coupled pendulums which swing to and fro in planes which are perpendicular to the plane of equilibrium.

equilibrium. The resulting motions of the points Q and R cause the other pendulum to start swinging, and the key parameter governing this coupling may be shown to be

$$\beta = \frac{1}{(l/l_1 + 1)(d/d_1 + 2)}. \tag{5.33}$$

If d is large enough then β will be small, and the coupling will be weak. The first pendulum will then come virtually to rest after $1/2\beta$ oscillation cycles of period $2\pi\sqrt{(l+l_1)/g}$, and all its energy will have been transferred to the other pendulum, which will then be oscillating vigorously. As in the previous example, this process will then reverse, with a total to-and-fro exchange time of $1/\beta$ cycles, until, in practice, friction damps out the motion.

5.5 Nonlinear oscillations

We begin this concluding section with an entirely new look at the spring oscillator of Fig. 5.2, using the ideas of *phase space* which we introduced briefly in Section 3.6.

Consider first small-amplitude motions, so that the oscillator is governed by

$$\ddot{x} = -\omega^2 x \tag{5.34}$$

(see (5.12)).

Our first move is to let y denote $\dot{x}$, the velocity of the particle in Fig. 5.2, and recast (5.34) as a *pair* of *first*-order equations:

$$\dot{x} = y, \tag{5.35a}$$

$$\dot{y} = -\omega^2 x \tag{5.35b}$$

(see Section 3.5 and (3.52)).

Next we use the chain rule to eliminate t, writing

$$\frac{\mathrm{d}y}{\mathrm{d}x} = \frac{\dot{y}}{\dot{x}} = -\frac{\omega^2 x}{y}, \tag{5.36}$$

so that

$$y\frac{\mathrm{d}y}{\mathrm{d}x} + \omega^2 x = 0. \tag{5.37}$$

We may integrate this immediately with respect to x to obtain

$$\tfrac{1}{2}y^2 + \tfrac{1}{2}\omega^2 x^2 = \text{constant}, \tag{5.38}$$

and these curves form a set of ellipses in the **phase plane** with coordinates x, y (Fig. 5.14). The arrows on these **phase paths** have been appended by reference to (5.35a); x clearly increases with time when $y > 0$ and decreases with time when $y < 0$.

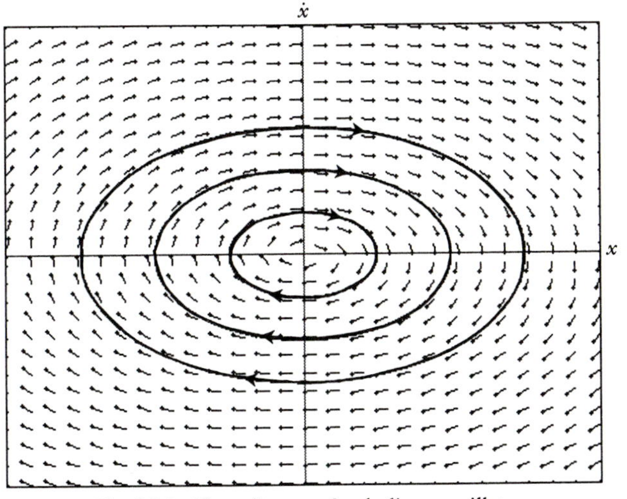

Fig. 5.14 Phase diagram for the linear oscillator.

Now, the *state* of the system in Fig. 5.2 is specified at any time by the position of the mass, x, together with its velocity $y = \dot{x}$ at that time; it therefore corresponds to a single point with coordinates (x, y) in phase space. Clearly, then, each closed curve in the phase diagram corresponds to a periodic motion of the actual system (Fig. 5.2), because the sequence of states keeps repeating itself each time a phase point (x, y) travels round the closed curve in question.

The program 2XTPHASE* is convenient for gaining familiarity with the phase plane. Using it, we may integrate (5.35)—or any other pair of autonomous first-order equations—by the Euler or improved Euler methods (Sections 4.4, 4.5) and then see the solution evolve simultaneously on the x-t graph and in the phase plane. Figure 5.15 shows one such example, and the lettering *ABCDA* has been added to highlight the corresponding points of the two representations.

For this case of the linear oscillator (5.34), of course, we already know the general solution (5.13), and

$$x = x_0 \cos \omega t \tag{5.39}$$

is therefore the solution satisfying the initial conditions $x = x_0$, $\dot{x} = 0$ at $t = 0$. This emerges, too, from our current formulation if we set $y = \dot{x}$ in (5.38), apply the initial conditions, and solve the resulting first-order equation for x as a function of t (see Ex. 3.6).

More generally, however, an explicit solution akin to (5.39) is not possible,

* See Appendix B.

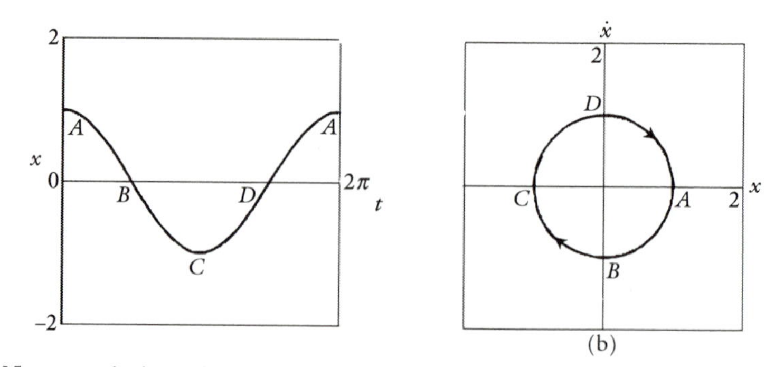

Fig. 5.15 x-t and phase plane plots of a simple harmonic oscillation, with $\omega=1$ and initial conditions $x=1$, $\dot{x}=0$ at $t=0$.

and the insight obtained from the phase diagram—which may still be constructed—is all the more valuable. A good example of this is provided, once again, by the simple pendulum.

Large swings of the pendulum

The simple pendulum equation

$$\frac{d^2\theta}{dt^2} + \frac{g}{l}\sin\theta = 0 \tag{5.40}$$

is nonlinear, and in Section 5.1 we gave only an approximate (linearized) treatment of small oscillations. Here we shall consider oscillations of any amplitude and, indeed, motions in which the pendulum whirls continually about the pivot. We shall take

$$\theta = \theta_0, \qquad \frac{d\theta}{dt} = \Omega \qquad \text{at } t = 0, \tag{5.41}$$

where θ_0 and Ω are given constants.

It is convenient to first change to the dimensionless time variable

$$\tilde{t} = t/(l/g)^{1/2}, \tag{5.42}$$

which corresponds to measuring time in units of $(l/g)^{1/2}$ (see Section 4.4). The problem then becomes

$$\ddot{\theta} + \sin\theta = 0, \tag{5.43}$$

with

$$\theta = \theta_0, \qquad \dot{\theta} = \tilde{\Omega} \qquad \text{at } \tilde{t} = 0, \tag{5.44}$$

where

$$\tilde{\Omega} = \Omega(l/g)^{1/2}, \tag{5.45}$$

and a dot denotes differentiation with respect to $\tilde{t}$.

Next, we recast (5.43) as a first-order system:

$$\dot{\theta} = y, \qquad \dot{y} = -\sin\theta, \tag{5.46a, b}$$

and integrate subject to $\theta = \theta_0$, $y = \tilde{\Omega}$ at $\tilde{t} = 0$ using the improved Euler method in the program 2XTPHASE.

Suppose first that $\tilde{\Omega} = 0$, so that the pendulum is released from rest. Then if θ_0 is small the result is an oscillation which is very nearly simple harmonic, in agreement with (5.5). Even if θ_0 is $\pi/2$ we see from Fig. 5.16(a) that the form of the oscillation is not much changed, with an almost circular path in the phase plane, and a period only some 18% larger than the small-amplitude value of $2\pi\sqrt{l/g}$, or 2π in dimensionless units.

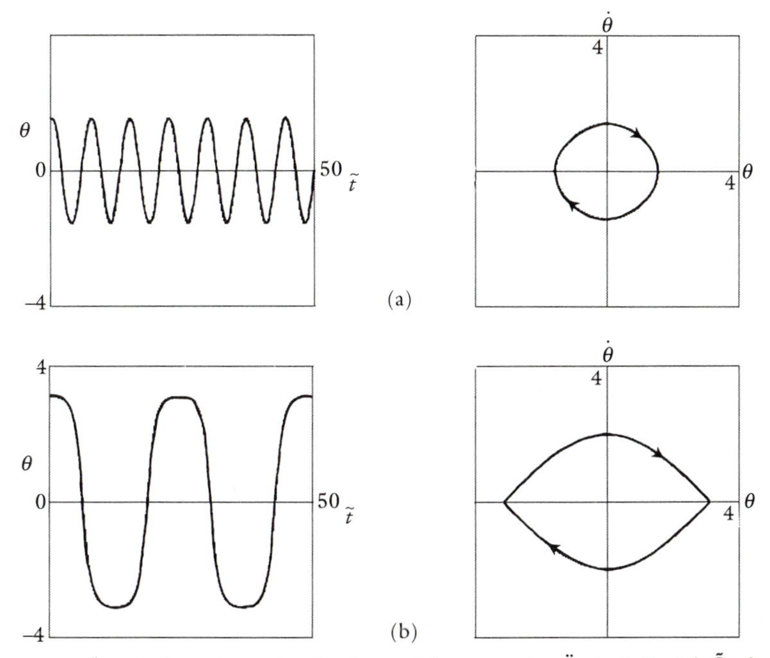

Fig. 5.16 θ-$\tilde{t}$ *and phase plane plots for the pendulum equation* $\ddot{\theta} + \sin\theta = 0$ *with* $\tilde{\Omega} = 0$ *and (a)* $\theta_0 = \pi/2$, *or 90° and (b)* $\theta_0 = 3.124139$, *or 179°.*

But if the pendulum is released from rest *very close to its 'upside-down' position* we see from Fig. 5.16(b) that the form of the oscillation is quite different, and the period is much longer. The period can, indeed, be made arbitrarily large by taking θ_0 sufficiently close to π, though increasing care must be taken in practice to avoid build-up of numerical errors (Ex. 5.4).

Suppose now that $\tilde{\Omega}$ is *not* necessarily zero, so that we may give the pendulum an angular velocity at the initial instant. In Fig. 5.17(a) we have used the program 2PHASE to generate a direction field in the phase plane and to trace out various phase paths starting from different initial conditions.

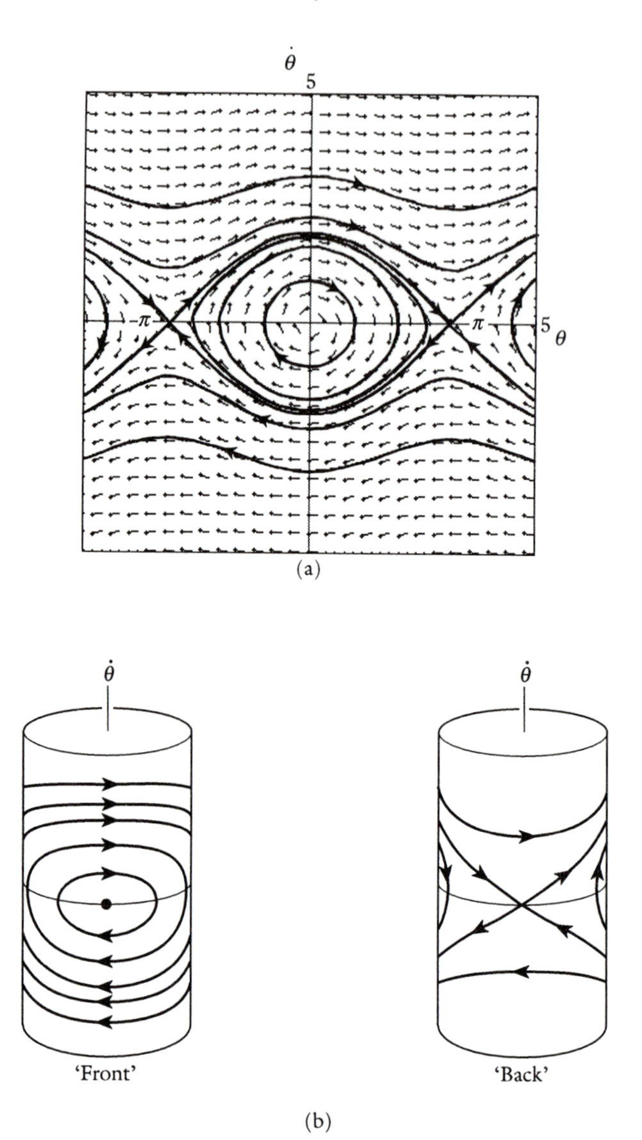

Fig. 5.17 (a) Phase plane for the simple pendulum, with the pattern repeating periodically as θ increases or decreases beyond the interval shown. (b) Paths in cylindrical phase space, which is more naturally suited to this problem.

If, for example, $\theta = 0$ at $t = 0$, then values of $\tilde{\Omega}$ less than 2 lead to closed curves around the origin in the phase plane, signifying to-and-fro swinging motions of the pendulum itself. But values of $\tilde{\Omega}$ greater than 2 lead to phase paths which leave the region shown, with $\dot{\theta}$ periodically increasing and decreasing, but never changing sign; these clearly correspond to persistent whirling motions of the pendulum about its pivot.

Because $\theta = \pi$ and $\theta = -\pi$ represent identical positions of the pendulum in space it is in fact best to wrap the portion of the phase plane $-\pi < \theta < \pi$ round on itself, and view the phase space as the surface of a cylinder, as in Fig. 5.17(b). All periodic motions are then represented by closed curves, the ones encircling the cylinder being 'whirling' ones, the others corresponding to motions in which the pendulum swings to-and-fro.

This whole idea of closed curves in phase space corresponding to *periodic* motions of the real, physical system is a very fundamental one, and we shall see more of it in Chapter 11.

Exercises

5.1 *The damped linear oscillator.* Use (3.31) to show that the general solution to (5.14) is

$$x = C\,\mathrm{e}^{-kt/2}\cos\left\{\left(\omega^2 - \tfrac{1}{4}k^2\right)^{1/2} t - D\right\},$$

provided $\frac{1}{4}k^2 < \omega^2$.

Confirm the main features of these decaying oscillations with the program 2XTPHASE from Appendix B.

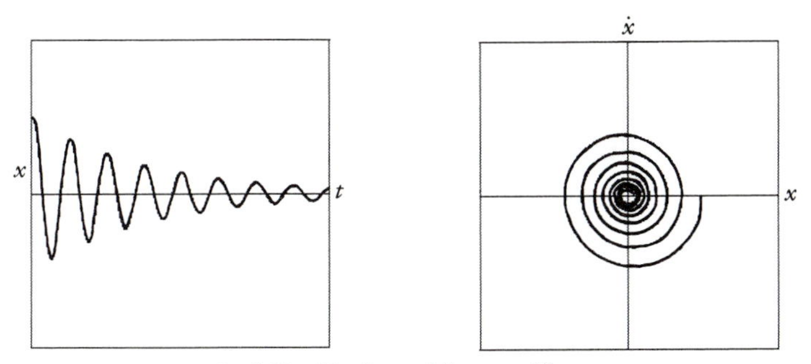

Fig. 5.18 The damped linear oscillator.

5.2 *Forced oscillations and resonance.* Consider the forced linear oscillator equation

$$\ddot{x} + \omega^2 x = a\cos\Omega t$$

(see (5.16)) subject to the initial conditions $x = \dot{x} = 0$ when $t = 0$.

Show that when $\Omega \neq \omega$ the solution is

$$x = \frac{a}{(\omega^2 - \Omega^2)}(\cos\Omega t - \cos\omega t),$$

and verify by direct substitution in the equation that in the resonant case, $\Omega = \omega$, the solution is

$$x = \frac{a}{2\omega} t \sin \omega t$$

instead.

5.3 *Multiple modes.* Extend the analysis in Section 5.3 to the case when the two masses m_1, m_2 in Fig. 5.8 are not equal. Obtain the counterpart of (5.22), solve the quadratic for ω^2, and show that if m_2 is much greater than m_1 then there is either a low-frequency mode with

$$\omega^2 \doteqdot \frac{3\alpha}{2m_2} \qquad \text{and} \qquad B \doteqdot 2A$$

or a high-frequency mode with

$$\omega^2 \doteqdot \frac{2\alpha}{m_1} \qquad \text{and} \qquad B \doteqdot -\frac{m_1}{2m_2} A.$$

5.4 *Large-amplitude pendulum oscillations.* Use the program 2XTPHASE to confirm the results in Fig. 5.16, with an oscillation period of $24.5(l/g)^{1/2}$ when $\theta_0 = 3.124139$, i.e. 179°.

What is the oscillation period when (a) $\theta_0 = 178°$, (b) $\theta_0 = 179.5°$?

5.5 *The whirling pendulum.* Suppose that a simple pendulum is hanging downward, and that we give it an initial angular velocity Ω, so that $\theta = 0$, $\mathrm{d}\theta/\mathrm{d}t = \Omega$ when $t = 0$. Show that the pendulum will only overshoot $\theta = \pi$ if

$$\Omega > 2(g/l)^{1/2}.$$

Suppose now that friction is present and proportional to $\mathrm{d}\theta/\mathrm{d}t$, so that in place of (5.40) we have

$$\frac{\mathrm{d}^2\theta}{\mathrm{d}t^2} + k\frac{\mathrm{d}\theta}{\mathrm{d}t} + \frac{g}{l}\sin\theta = 0.$$

If we write $\tilde{t} = t(g/l)^{1/2}$ we obtain

$$\ddot{\theta} + \tilde{k}\dot{\theta} + \sin\theta = 0$$

with $\dot{\theta} = \tilde{\Omega}$ at $\tilde{t} = 0$, where $\tilde{\Omega} = \Omega(l/g)^{1/2}$, $\tilde{k} = k(l/g)^{1/2}$ and a dot denotes differentiation with respect to $\tilde{t}$.

Use the program PENDANIM to solve this equation by a double-precision Runge–Kutta method and display a simple animation of the motion. For $\tilde{k} = 0.1$, say, how many complete revolutions about the pivot does the pendulum make when (i) $\Omega = 4(g/l)^{1/2}$, (ii) $\Omega = 10(g/l)^{1/2}$?

6 *Planetary motion*

6.1 Introduction

The planets move, very nearly, according to three simple rules:

1. The orbit of each planet is an ellipse, with the Sun at one focus.
2. A line drawn from the Sun to a planet sweeps out equal areas in equal times.
3. The periods of the various orbits are in proportion to $\bar{r}^{3/2}$, where $\bar{r}$ denotes the mean distance of a planet from the Sun.

These rules were discovered by Kepler after a painstaking analysis of the astronomical observations, and they appear, somewhat obscurely, in two major books of 1609 and 1619. The third rule turned out to be particularly important, for it helped point the way towards Newton's inverse-square law of gravitation.

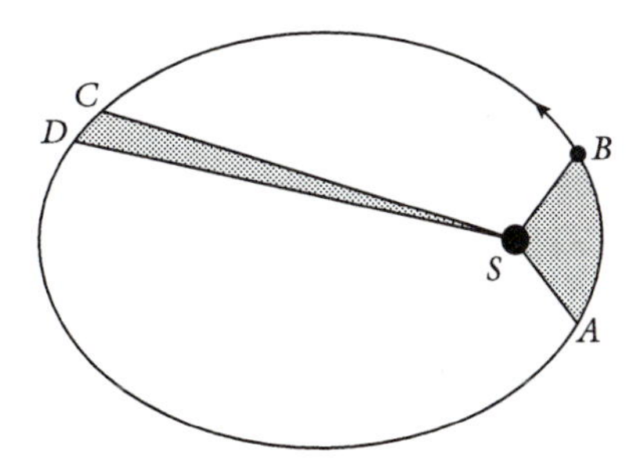

Fig. 6.1 Planetary motion about the Sun S. The planet takes as long to get from C to D as it does to get from A to B, the areas ABS and CDS being equal.

Before we see how this happened it may be helpful to say a little about the geometry of motion in an ellipse. With respect to suitably chosen axes an ellipse has equation

$$\frac{X^2}{a^2} + \frac{Y^2}{b^2} = 1, \tag{6.1}$$

where we shall take $b \leq a$ without loss of generality. The transformation $X = aX'$, $Y = bY'$ yields a circle, so an ellipse is a circle which has been 'squashed' in one direction. The smaller the ratio b/a the more 'squashed' it is, and a convenient measure of this squashing is the **eccentricity**

$$e = (1 - b^2/a^2)^{1/2}, \tag{6.2}$$

which lies between 0 and 1.

The **focal points** of the ellipse are at

$$F, F' = (\pm ae, 0), \tag{6.3}$$

so the more eccentric the ellipse the more off-centre are the focal points. These are significant in two ways. First, the lines $F'P$ and FP in Fig. 6.2 make equal angles with the tangent to the ellipse at P, so that an elliptical mirror reflects a ray of light from one focus to the other. Second, as the point P moves around the ellipse the distance $F'P + PF$ remains constant, and equal to $2a$.

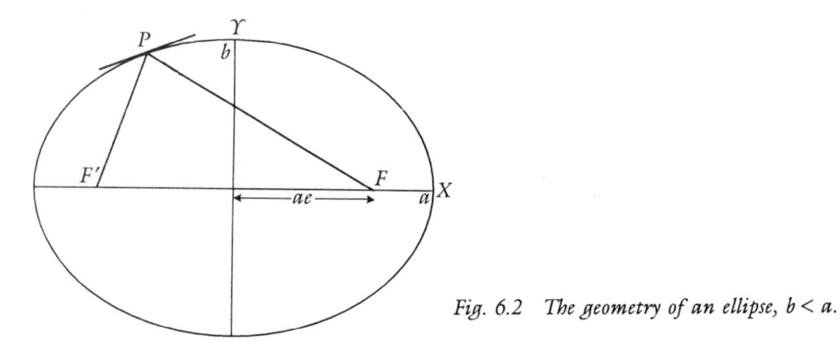

Fig. 6.2 The geometry of an ellipse, $b < a$.

Let us return now to the dynamical problem, and note first that while the planets do indeed move in ellipses with the Sun at one focus, the eccentricities of these orbits are very small (Table 6.1), so the orbits are *almost circular*. Further, when a body of mass m moves in a circular orbit of radius r with speed v, it has at any instant an acceleration v^2/r towards the centre of the circle, without which it would continue in a straight line along a tangent (Fig. 6.3). This therefore requires a

$$\text{centripetal force} = \frac{mv^2}{r} \tag{6.4}$$

acting continually on the body, towards the centre of the circle.

Now, the period of a circular orbit is $T = 2\pi r/v$, and if this is proportional to $r^{3/2}$, from Kepler's third rule, then v must be proportional to $r^{-1/2}$. This implies in turn that the centripetal acceleration v^2/r is proportional to $1/r^2$.

It was an argument of this general kind, apparently, which led Newton and his contemporaries to the idea that the Sun attracts the planets according to an inverse-square law. This is still a long way, of course, from Newton's

Table 6.1 Planetary orbit data. The first six planets were known in Kepler's time, while the others were discovered later.

	$\bar{r}$ (units of $\bar{r}_{\text{Earth}}$)	Period (years)	Eccentricity of elliptical orbit	Mass (units of M_{Earth})
Mercury	0.387	0.241	0.206	0.055
Venus	0.723	0.615	0.007	0.815
Earth	1.000	1.000	0.017	1.000
Mars	1.524	1.881	0.093	0.107
Jupiter	5.203	11.862	0.048	317.94
Saturn	9.539	29.46	0.056	95.18
Uranus (1781)	19.191	84.02	0.046	14.53
Neptune (1846)	30.061	164.77	0.010	17.13
Pluto (1930)	39.529	247.68	0.248	0.0022

$\bar{r}_{\text{Earth}} = 1.50 \times 10^{11}$ m $\quad M_{\text{Earth}} = 5.97 \times 10^{24}$ kg
$G = 6.67 \times 10^{-11}$ m^3/kg·s^2 $\quad M_{\text{Sun}} = 1.99 \times 10^{30}$ kg

eventual idea of *universal* gravitation, which is that *any* two point masses m_1 and m_2, distant r apart, attract one another with a force

$$F = \frac{Gm_1m_2}{r^2}, \tag{6.5}$$

where G is a universal constant. Nonetheless, it led immediately to the big

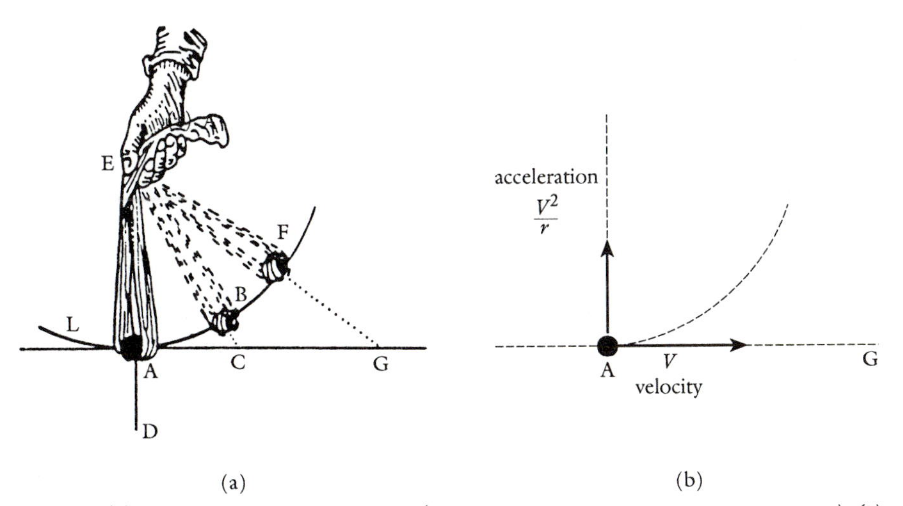

Fig. 6.3 (a) A stone being whirled in a sling (from Descartes' Principles of Philosophy, *1644). (b) The velocity and acceleration of the stone, at the instant at which it is at the point A.*

question of the day: is an inverse-square law of attraction towards the Sun consistent with the fact that planetary orbits are actually *elliptical* in shape?

6.2 Equations of motion under a central force

Suppose that a particle P of mass m moves in a certain plane under a force which is always directed towards some fixed point O. Rather than assume from the outset an inverse-square law of force, we shall begin with a more general problem, letting the force be $f(r)$, where r denotes the distance OP (Fig. 6.4).

In order to obtain the differential equations of motion it is convenient to use the coordinates (r, θ) rather than (x, y), and we first resolve both the velocity and the acceleration at any instant into (i) a 'radial' component, parallel to OP, and (ii) a 'transverse' component, perpendicular to OP.

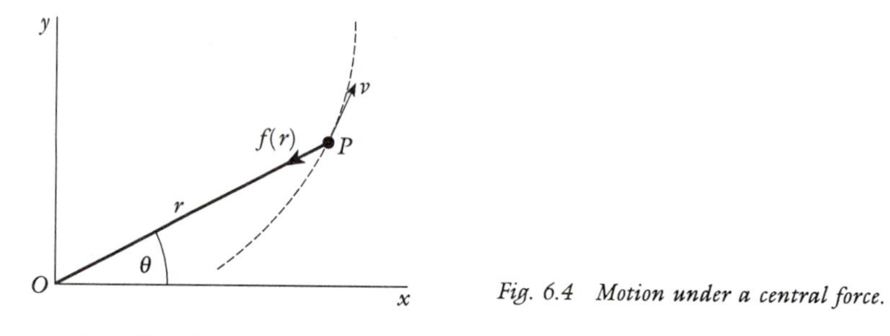

Fig. 6.4 Motion under a central force.

One quite effective way of doing this is to imagine for a moment that the events in Fig. 6.4 are taking place in the *complex* plane, i.e. the Argand diagram. The particle's 'complex position' at any instant is then $z = x + \mathrm{i}y$, or

$$z = r\,\mathrm{e}^{\mathrm{i}\theta} \tag{6.6}$$

(see (2.33)). Differentiating with respect to t we obtain

$$\begin{aligned}\dot{z} &= \dot{r}\,\mathrm{e}^{\mathrm{i}\theta} + r\,\mathrm{i}\dot{\theta}\,\mathrm{e}^{\mathrm{i}\theta} \\ &= \dot{r}\,\mathrm{e}^{\mathrm{i}\theta} + r\dot{\theta}\,\mathrm{e}^{\mathrm{i}(\theta+\pi/2)},\end{aligned} \tag{6.7}$$

so that the radial and transverse components of *velocity* are

$$\dot{r}, r\dot{\theta} \tag{6.8}$$

(Fig. 6.5(a)). By differentiating (6.7) in turn, and after a little rearrangement, we find that

$$\ddot{z} = (\ddot{r} - r\dot{\theta}^2)\mathrm{e}^{\mathrm{i}\theta} + (2\dot{r}\dot{\theta} + r\ddot{\theta})\mathrm{e}^{\mathrm{i}(\theta+\pi/2)}, \tag{6.9}$$

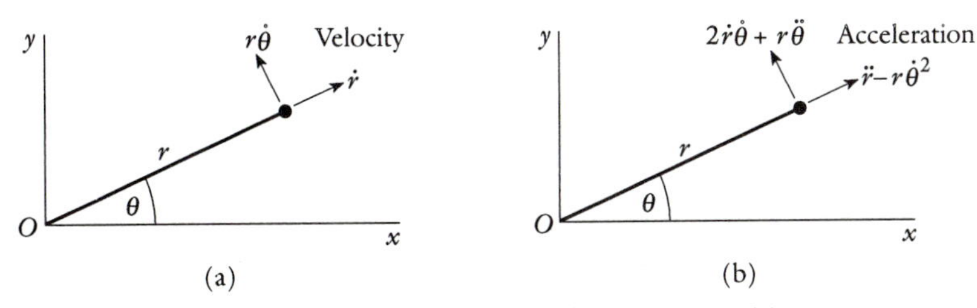

Fig. 6.5 Radial and transverse components of (a) velocity and (b) acceleration.

so that the radial and transverse components of *acceleration* are

$$\ddot{r}-r\dot{\theta}^2, 2\dot{r}\dot{\theta}+r\ddot{\theta} \tag{6.10}$$

(Fig. 6.5(b)).

We now apply the fundamental law of mechanics, 'force = mass × acceleration'. As there is a radial component of force $f(r)$ *towards* O in Fig. 6.4, but no transverse component, we obtain

$$\begin{aligned} m(\ddot{r}-r\dot{\theta}^2) &= -f(r), \\ m(2\dot{r}\dot{\theta}+r\ddot{\theta}) &= 0 \end{aligned} \tag{6.11a, b}$$

as our *differential equations of motion.*

As a simple check on these equations we note briefly that in the special case of motion in a circle about O with radius r and speed v we have $\dot{r}=\ddot{r}=0$ and $r\dot{\theta}=v$, by (6.8). Equation (6.11a) then reduces to $mv^2/r=f(r)$, in keeping with (6.4).

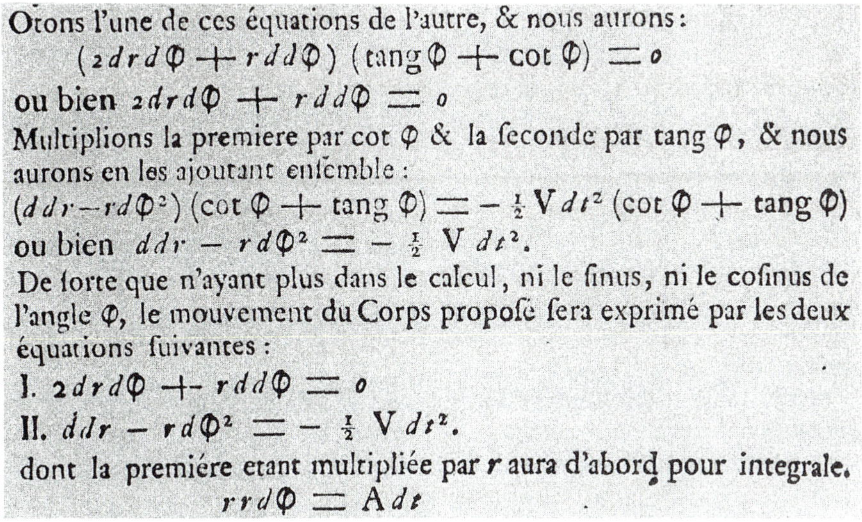

Otons l'une de ces équations de l'autre, & nous aurons:

$$(2dr d\Phi + r dd\Phi)(\tang\Phi + \cot\Phi) = 0$$

ou bien $2dr d\Phi + r dd\Phi = 0$

Multiplions la premiere par cot Φ & la ſeconde par tang Φ, & nous aurons en les ajoutant enſemble :

$$(ddr - r d\Phi^2)(\cot\Phi + \tang\Phi) = -\tfrac{1}{2}V dt^2 (\cot\Phi + \tang\Phi)$$

ou bien $ddr - r d\Phi^2 = -\tfrac{1}{2} V dt^2$.

De ſorte que n'ayant plus dans le calcul, ni le ſinus, ni le coſinus de l'angle Φ, le mouvement du Corps propoſé ſera exprimé par les deux équations ſuivantes :

I. $2dr d\Phi + r dd\Phi = 0$

II. $ddr - r d\Phi^2 = -\tfrac{1}{2} V dt^2$.

dont la premiére etant multipliée par r aura d'abord pour integrale.

$$rr d\Phi = A dt$$

Fig. 6.6 The first appearance of eqns. 6.11(a, b), in a paper by Euler published in 1749.

6.3 The equal-area rule

We shall in due course be most interested in the case of an inverse-square law, $f(r) = c/r^2$, but we note first a result which holds for *any* central force $f(r)$, inverse square or otherwise.

The result follows immediately from (6.11b), for on multiplying by r we obtain

$$2r\dot{r}\dot{\theta} + r^2\ddot{\theta} = 0,$$

and the new left-hand side can be recognized as the derivative of the product $r^2\dot{\theta}$, so

$$r^2\dot{\theta} = \text{constant}. \tag{6.12}$$

The angular velocity $\dot{\theta}$ is therefore large when r is small, and vice versa.

Equation (6.12) also has a simple interpretation in terms of the rate at which the line OP sweeps out area. We see from Fig. 6.7 that as the particle

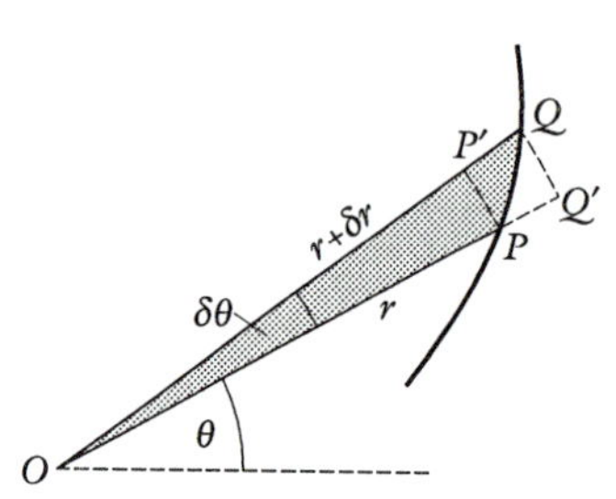

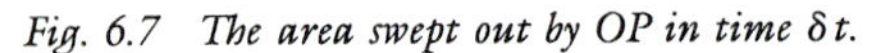
Fig. 6.7 The area swept out by OP in time δt.

moves from P to Q in time δt, the line OP sweeps out an area δA such that

$$\tfrac{1}{2}r^2\,\delta\theta < \delta A < \tfrac{1}{2}(r + \delta r)^2\,\delta\theta,$$

the lower and upper bounds here being the areas of the circular sectors OPP' and OQQ' respectively. On dividing by δt and letting $\delta t \to 0$ (so that $\delta r \to 0$ and $\delta\theta \to 0$ also), we find $\delta A/\delta t$ sandwiched between two quantities which both tend to $\frac{1}{2}r^2\dot{\theta}$. Thus

$$\frac{\mathrm{d}A}{\mathrm{d}t} = \tfrac{1}{2}r^2\dot{\theta}, \tag{6.13}$$

and (6.12) therefore implies that a planet moving under the action of a central force does so in such a way that *the line OP sweeps out area at a constant rate.*

In this way, then, Kepler's second rule is simply a consequence of the gravitational force on a planet being *central*, i.e. directed towards the Sun. This result was first established by Newton, though not using the methods above. As Fig. 6.8 shows, Newton does not treat directly a continuously acting

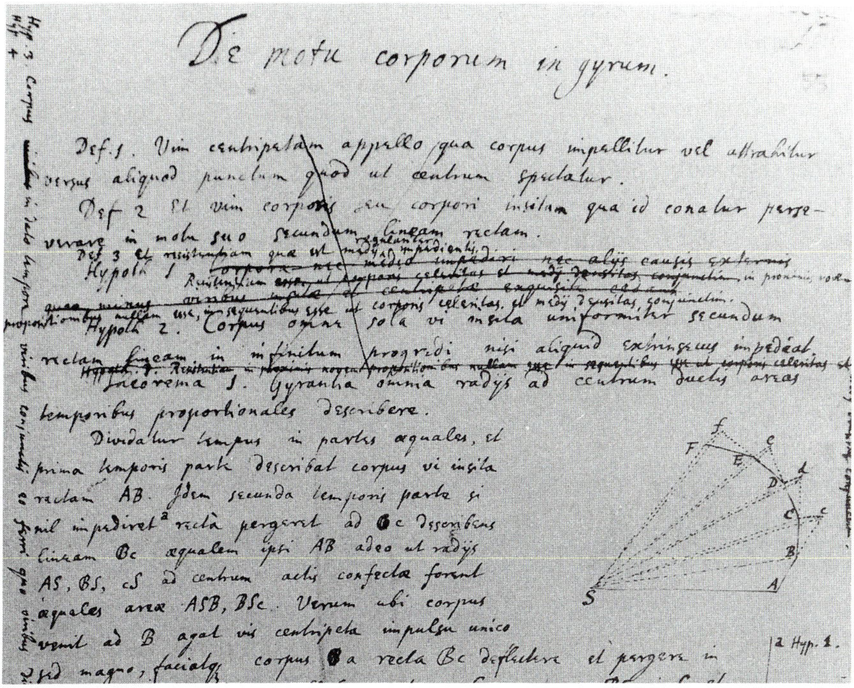

Fig. 6.8 Part of Newton's initial draft of De motu corporum in gyrum, *a short paper sent to Halley in 1684, containing in particular a dynamical argument leading to Kepler's 2nd rule of planetary motion. (Cambridge University Library).*

force at all, but replaces it instead by a succession of impulses towards S; these give sudden deflections (at B, C, D, etc.) to a motion which between times takes place uniformly in a straight line.

6.4 Differential equation for the orbit

We now consider the *path* taken by a particle moving under a central force $f(r)$. We therefore want a differential equation for r not as a function of t but as a function of θ. To this end, we start with (6.11a):

$$m(\ddot{r} - r\dot{\theta}^2) = -f(r) \tag{6.14}$$

and then use (6.12), i.e.

$$r^2\dot{\theta} = \hbar, \tag{6.15}$$

say, to convert derivatives with respect to t into derivatives with respect to θ.

Note first that

$$\frac{\mathrm{d}r}{\mathrm{d}t} = \frac{\mathrm{d}r}{\mathrm{d}\theta}\frac{\mathrm{d}\theta}{\mathrm{d}t} = \frac{\hbar}{r^2}\frac{\mathrm{d}r}{\mathrm{d}\theta} = -\hbar\frac{\mathrm{d}}{\mathrm{d}\theta}\left(\frac{1}{r}\right).$$

This gives us a hint that it may be more convenient to work instead with the new variable

$$u = \frac{1}{r} \tag{6.16}$$

rather than with r itself. In terms of the variables u, θ we then have

$$\begin{aligned} &\text{radial velocity component} \qquad \dot{r} = -h\frac{du}{d\theta}, \\ &\text{transverse velocity component} \qquad r\dot{\theta} = hu \end{aligned} \tag{6.17a, b}$$

(see (6.8)).

Moreover,

$$\ddot{r} = \frac{d\dot{r}}{dt} = \frac{d\dot{r}}{d\theta}\frac{d\theta}{dt} = -h^2u^2\frac{d^2u}{d\theta^2},$$

and on substituting into (6.14) we find, with a little rearrangement,

$$\frac{d^2u}{d\theta^2} + u = \frac{f\left(\frac{1}{u}\right)}{mh^2u^2} \tag{6.18}$$

as the differential equation for the path of a particle of mass m moving under a central force $f(r)$.

In general, this equation is difficult to solve, because the right-hand side will typically be some nonlinear function of the dependent variable u. Happily, however, it may be solved very easily in the case of most practical interest, namely $f(r) \propto 1/r^2$.

6.5 Orbits under an inverse-square law

Suppose now that the point mass m moves under the gravitational attraction of a point mass M which is fixed at O. According to (6.5), then,

$$f(r) = \frac{GMm}{r^2}. \tag{6.19}$$

We take as initial conditions

$$\left.\begin{aligned} r = d, \quad \theta = 0 \\ \dot{r} = 0, \quad r\dot{\theta} = v \end{aligned}\right\} \quad \text{when } t = 0 \tag{6.20}$$

(Fig. 6.9). Using (6.16) and (6.17a) these imply that

$$u = \frac{1}{d}, \qquad \frac{du}{d\theta} = 0 \qquad \text{when } \theta = 0 \tag{6.21}$$

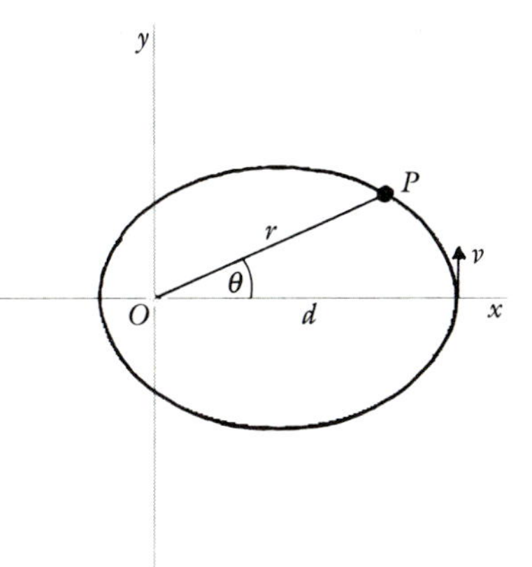

Fig. 6.9 Elliptical orbit under an inverse-square law of force, $v < v_c$. (See also Fig. 6.11.)

and they also tell us the value of the constant h in (6.15), for $r = d$ and $r\dot{\theta} = v$ when $\theta = 0$,

$$h = vd. \tag{6.22}$$

Using (6.19), i.e. $f(1/u) = GMmu^2$, the orbit equation (6.18) then reduces to

$$\frac{\mathrm{d}^2 u}{\mathrm{d}\theta^2} + u = \frac{GM}{d^2 v^2}. \tag{6.23}$$

Most conveniently, the right-hand side is independent of u, so we have a linear differential equation with constant coefficients. The general solution is accordingly

$$u = A \cos\theta + B \sin\theta + \frac{GM}{d^2 v^2} \tag{6.24}$$

(see Section 3.4). We determine the arbitrary constants A and B by applying the initial conditions (6.21), obtaining

$$u = \left(\frac{1}{d} - \frac{GM}{d^2 v^2}\right)\cos\theta + \frac{GM}{d^2 v^2}. \tag{6.25}$$

If we define

$$v_c = \left(\frac{GM}{d}\right)^{1/2} \tag{6.26}$$

and recall that $u = 1/r$ we may rewrite (6.25) as

$$r = \frac{d\dfrac{v^2}{v_c^2}}{1 + \left(\dfrac{v^2}{v_c^2} - 1\right)\cos\theta}. \tag{6.27}$$

The significance, then, of the velocity v_c defined by (6.26) is that if $v = v_c$ the particle moves in a circle about O, with $r = d$.

To explore (6.27) more generally, define

$$e_* = \frac{v^2}{v_c^2} - 1 \tag{6.28}$$

and

$$a = \frac{d}{1 - e_*}, \qquad e_* \neq 1. \tag{6.29}$$

With a little algebra, (6.27) can then be cast into the form

$$\frac{(x + ae_*)^2}{a^2} + \frac{y^2}{a^2(1 - e_*^2)} = 1. \tag{6.30}$$

On comparing this with (6.1) we see that the path is an *ellipse*, centre $(-ae_*, 0)$, if $|e_*| < 1$, i.e. if

$$\frac{v}{v_c} < \sqrt{2}. \tag{6.31}$$

Elliptical orbits *are*, therefore, consistent with an inverse square law of attraction, as we had hoped.

Moreover, the origin O in Fig. 6.9, where the attracting mass M is located, is at one *focus* of the ellipse (6.30). To see this, note that $b^2 = a^2(1 - e_*^2)$ in this case, so by (6.2) the eccentricity of the ellipse is $e = |e_*|$. The centre of

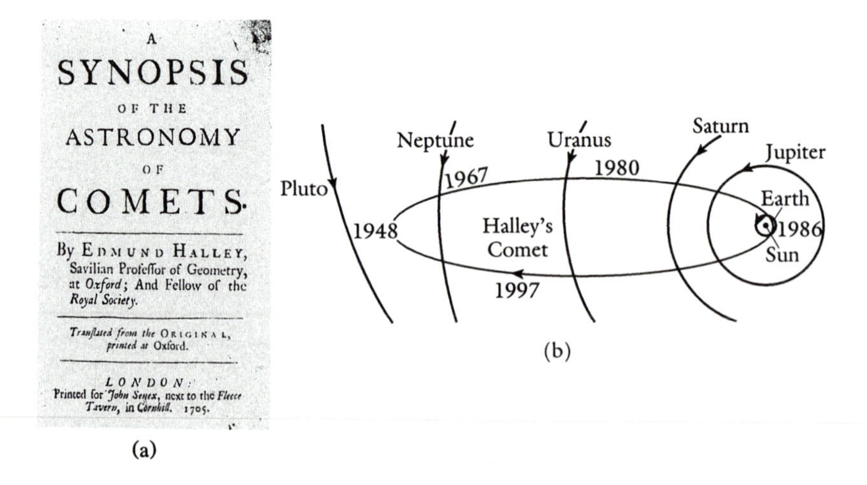

A
SYNOPSIS
OF THE
ASTRONOMY
OF
COMETS.

By EDMUND HALLEY,
Savilian Profeſſor of Geometry,
at *Oxford*; And Fellow of the
Royal Society.

Tranſlated from the ORIGINAL,
printed at Oxford.

LONDON:
Printed for *John Senex*, next to the *Fleece Tavern*, in *Cornhill*. 1705.

Fig. 6.10 (a) Halley's book on comets, 1705. (b) The orbit of comet Halley.

the ellipse is therefore at a distance ae from the origin in Fig. 6.9, so the origin there must be a focus, by Fig. 6.2.

We may find the **period** T of the orbit by noting from (6.13) that the line OP in Fig. 6.9 is sweeping out area at a rate $\frac{1}{2}h$, i.e. $\frac{1}{2}vd$. As the area of the ellipse (6.1) is πab it follows that

$$T = \frac{\pi ab}{\frac{1}{2}vd} = 2\pi\left(\frac{a^3}{GM}\right)^{1/2}, \tag{6.32}$$

so $T \propto a^{3/2}$, which is consistent with Kepler's third rule of planetary motion.

Another pleasing application of this last result is to Halley's comet, which has a far more elliptical orbit than any of the planets (Fig. 6.10). At its greatest distance from the Sun it is about 36 times further away than the Earth, so $2a = 36\bar{r}_{\text{Earth}}$, approximately. On using (6.32) and Table 6.1 we then obtain an orbital period of approximately 76 years for Halley's comet, as observed.

6.6 A numerical approach

We may of course tackle the whole problem in Section 6.5 by direct numerical integration instead. It is then more straightforward to use rectangular Cartesian coordinates x, y and to resolve the gravitational force towards the origin (6.19) into a component $GMm\cos\theta/r^2 = GMmx/r^3$ in the negative x-direction and a component $GMmy/r^3$ in the negative y-direction. The differential equations of motion are therefore

$$\frac{\mathrm{d}^2x}{\mathrm{d}t^2} = -\frac{GMx}{(x^2+y^2)^{3/2}},$$
$$\frac{\mathrm{d}^2y}{\mathrm{d}t^2} = -\frac{GMy}{(x^2+y^2)^{3/2}}, \tag{6.33a, b}$$

and the initial conditions corresponding to (6.20) will be

$$\left.\begin{array}{ll} x = d, & y = 0 \\ \dfrac{\mathrm{d}x}{\mathrm{d}t} = 0, & \dfrac{\mathrm{d}y}{\mathrm{d}t} = v \end{array}\right\} \quad \text{when } t = 0. \tag{6.34}$$

We first introduce non-dimensional variables to remove any unnecessary parameters from the computation (cf. Sections 4.4 and 5.5). We choose

$$\tilde{x} = \frac{x}{d}, \qquad \tilde{y} = \frac{y}{d}, \qquad \tilde{t} = \left(\frac{GM}{d^3}\right)^{1/2} t; \tag{6.35}$$

this is equivalent to measuring distance in units of d (see Fig. 6.9) and time in units of $(d^3/GM)^{1/2}$ (cf. (6.32)). In consequence, velocities are measured in units of $(GM/d)^{1/2}$ (cf. (6.26)).

In this non-dimensional formulation the equations become

$$\ddot{\tilde{x}} = -\frac{\tilde{x}}{(\tilde{x}^2+\tilde{y}^2)^{3/2}},\qquad \ddot{\tilde{y}} = -\frac{\tilde{y}}{(\tilde{x}^2+\tilde{y}^2)^{3/2}}, \tag{6.36a, b}$$

where a dot denotes differentiation with respect to $\tilde{t}$. In a similar way, the initial conditions (6.34) become

$$\left.\begin{array}{ll}\tilde{x}=1, & \tilde{y}=0\\ \dot{\tilde{x}}=0, & \dot{\tilde{y}}=\tilde{v}\end{array}\right\}\quad \text{when}\quad \tilde{t}=0, \tag{6.37}$$

where $\tilde{v}=v/v_c$, v_c being the special initial velocity (namely $(GM/d)^{1/2}$) which leads to a circular orbit, centre O (see (6.26)).

The final step is to recast (6.36) as a first-order system, by letting, say, $x_1=\tilde{x}$, $x_2=\tilde{y}$, $x_3=\dot{\tilde{x}}$, $x_4=\dot{\tilde{y}}$, so that

$$\begin{aligned}\dot{x}_1 &= x_3,\\ \dot{x}_2 &= x_4,\\ \dot{x}_3 &= -\frac{x_1}{(x_1^2+x_2^2)^{3/2}},\\ \dot{x}_4 &= -\frac{x_2}{(x_1^2+x_2^2)^{3/2}}.\end{aligned} \tag{6.38}$$

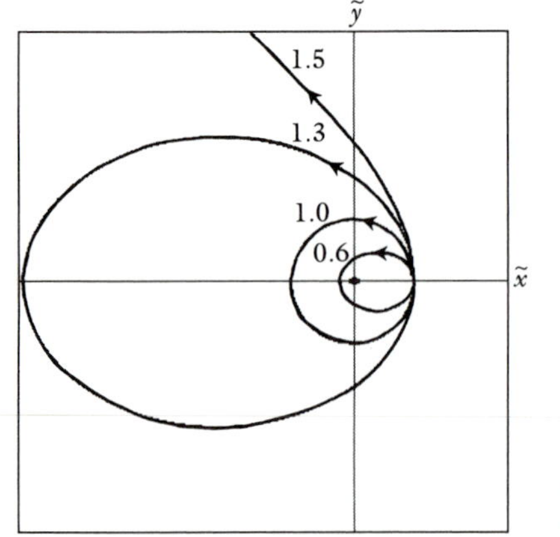

Fig. 6.11 Paths under an inverse-square law central force for 4 different values of v/v_c.

The program NPHASE in Appendix B can be adapted to integrate these equations by the Runge–Kutta method (see p. 216), and the resulting paths for four different values of the (dimensionless) initial velocity $\tilde{v}$ are shown in Fig. 6.11.

For $\tilde{v} < 1$ the path is an ellipse with the origin at the left-hand focus, for $\tilde{v} = 1$ the path is a circle about the origin, and for $\tilde{v} > 1$ the path is an ellipse with the origin at the right-hand focus, provided that $\tilde{v} < \sqrt{2}$. For $\tilde{v} > \sqrt{2}$ the path is no longer closed (it is, in fact, a hyperbola), and the particle flies off to infinity. All this can, of course, be deduced from the analysis in Section 6.5 (see, in particular, (6.28), (6.30) and (6.31)).

In addition to our standard check on the computation (halving the time step h) we may use *conservation of energy* in the present system. To see this, multiply (6.33a) by $\mathrm{d}x/\mathrm{d}t$, (6.33b) by $\mathrm{d}y/\mathrm{d}t$, and add. We may then integrate to obtain

$$\tfrac{1}{2}m\left\{\left(\frac{\mathrm{d}x}{\mathrm{d}t}\right)^2 + \left(\frac{\mathrm{d}y}{\mathrm{d}t}\right)^2\right\} - \frac{GMm}{r} = \text{constant}, \tag{6.39}$$

or, in dimensionless terms,

$$\tfrac{1}{2}\left\{(\dot{\tilde{x}})^2 + (\dot{\tilde{y}})^2\right\} - \frac{1}{(\tilde{x}^2 + \tilde{y}^2)^{1/2}} = \text{constant}. \tag{6.40}$$

The first term in (6.39) is the kinetic energy of the particle, while the second term, $-GMm/r$, is the potential energy of the mass m due to the gravitational field of the mass M. Their sum should be constant, so monitoring the left-hand side of (6.40) provides an additional check as the computation proceeds.

In view of (6.40), the 'planet' travels fastest when it is close to the 'Sun', i.e. the origin, and this can make the step-by-step method so inaccurate there that it leads to totally spurious 'results'. To overcome this problem it is best to introduce a *variable time step*, so that h automatically becomes smaller when r does (Ex. 6.3). This is a valid procedure with any of our three step-by-step methods, provided of course that the same h is used during any single up-dating of all the variables involved.

But what we have done so far amounts really only to a test of NPHASE and the Runge–Kutta routine, for we can in any case calculate the orbit of a single 'planet' by the method of Section 6.5. The real value of the computational approach in the present context comes when we consider the motion of *several* point masses attracting one another according to an inverse-square law (Section 6.8), and we take a step towards that in the next section.

6.7 The two-body problem

Suppose we have two attracting masses m_1, m_2 which are initially a distance d apart. Let

$$M = m_1 + m_2 \tag{6.41}$$

denote their combined mass. Then if we give the two masses initial velocities

$$v_1 = \frac{m_2}{M}\left(\frac{GM}{d}\right)^{1/2}, \qquad v_2 = \frac{m_1}{M}\left(\frac{GM}{d}\right)^{1/2} \tag{6.42a,b}$$

in the appropriate directions they will move in concentric circles about their *centre of mass* C, namely the point along the line joining them which divides that line in the ratio $m_2 : m_1$ (Fig. 6.12). Their common angular velocity will be

$$\Omega = \left(\frac{GM}{d^3}\right)^{1/2}, \tag{6.43}$$

(Ex. 6.4).

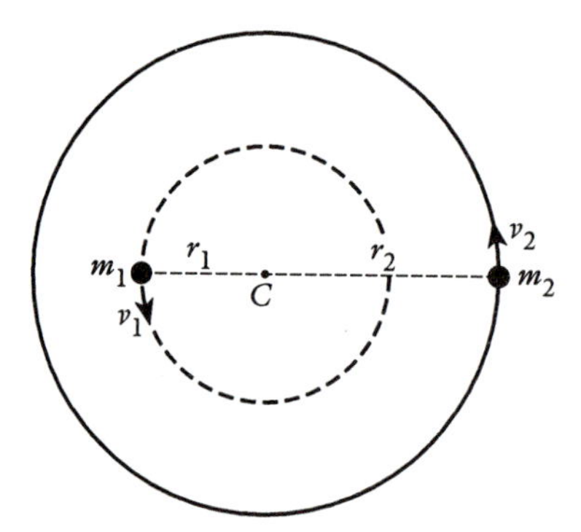

Fig. 6.12 Two attracting particles moving in concentric circles; $r_1/r_2 = m_2/m_1$.

This is, of course, a very special case of two-body motion; if v_1 and v_2 in (6.42) are both reduced by, say, a factor of 2, then the masses revolve instead in ellipses with C as a common focus.

A typical outcome of more general initial conditions still is shown in Fig. 6.13(a); the masses revolve around one another while drifting together, as a pair, through space. In fact, their centre of mass C drifts at a constant velocity, and relative to an observer moving with C their orbits are, again, simply ellipses (Fig. 6.13(b)).

In this general way, then, the motions of *two* attracting point masses are really quite simple.

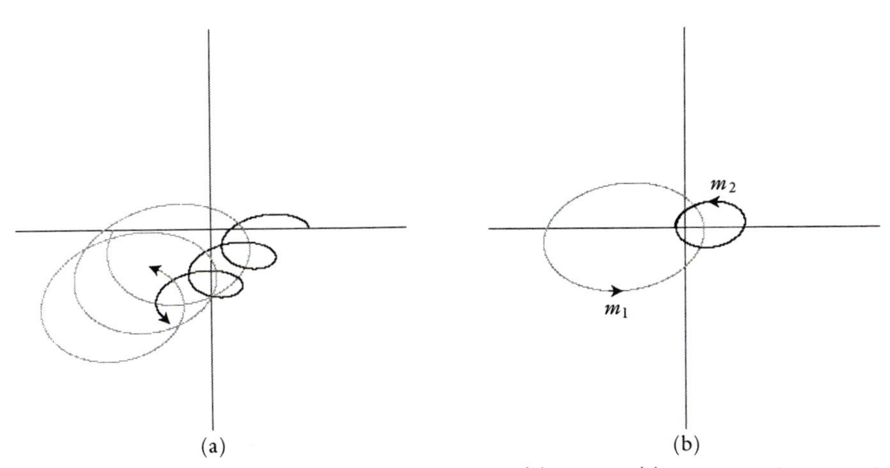

Fig. 6.13 Typical motion of two attracting point masses (a) in space (b) relative to the centre of mass C. In the case shown, $m_1/m_2=3/7$.

6.8 The three-body problem

As soon as *three* attracting bodies are involved, motions of extraordinary complexity can occur.

Let the masses be m_1, m_2, and m_3. The equations of motion for m_1 are now

$$m_1 \frac{d^2x_1}{dt^2} = \frac{Gm_1m_2}{r_{12}^3}(x_2 - x_1) + \frac{Gm_1m_3}{r_{31}^3}(x_3 - x_1),$$

$$m_1 \frac{d^2y_1}{dt^2} = \frac{Gm_1m_2}{r_{12}^3}(y_2 - y_1) + \frac{Gm_1m_3}{r_{31}^3}(y_3 - y_1), \qquad (6.44a,b)$$

and the equations of motion for m_2 and m_3 are similar, giving us in total *six* coupled differential equations of second order. Here

$$r_{12} = \left[(x_2 - x_1)^2 + (y_2 - y_1)^2\right]^{1/2}, \qquad (6.45)$$

and r_{23}, r_{31} are defined similarly (Fig. 6.14).

The total energy of the system is

$$E = \sum_{i=1}^{3} \tfrac{1}{2}m_i\left\{\left(\frac{dx_i}{dt}\right)^2 + \left(\frac{dy_i}{dt}\right)^2\right\}$$
$$- \frac{Gm_1m_2}{r_{12}} - \frac{Gm_2m_3}{r_{23}} - \frac{Gm_3m_1}{r_{31}} \qquad (6.46)$$

(cf. (6.39)), and it can be shown from the basic equations (6.44) that E is a constant, as we would expect.

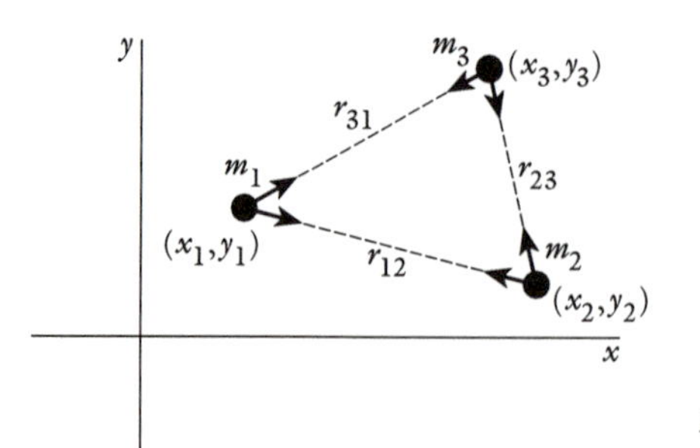

Fig. 6.14 *The 3-body problem.*

The *centre of mass C* is now defined as having coordinates

$$x_c = \frac{m_1x_1 + m_2x_2 + m_3x_3}{m_1 + m_2 + m_3}, \qquad y_c = \frac{m_1y_1 + m_2y_2 + m_3y_3}{m_1 + m_2 + m_3}, \tag{6.47}$$

and this again moves at a constant velocity determined by the initial conditions.

Non-dimensionalization

Let M denote a convenient unit of mass and d a convenient unit of length. We may then introduce the dimensionless quantities

$$\tilde{m}_i = \frac{m_i}{M}, \qquad \tilde{x}_i = \frac{x_i}{d}, \qquad \tilde{t} = \left(\frac{GM}{d^3}\right)^{1/2} t, \tag{6.48}$$

noting that our unit of velocity will therefore be $(GM/d)^{1/2}$.

One merit of the transformations (6.48) is that they remove the gravitational constant G from the governing equations (6.44), which now become

$$\ddot{\tilde{x}}_1 = \frac{\tilde{m}_2(\tilde{x}_2 - \tilde{x}_1)}{\tilde{r}_{12}^3} + \frac{\tilde{m}_3(\tilde{x}_3 - \tilde{x}_1)}{\tilde{r}_{31}^3},$$

$$\ddot{\tilde{y}}_1 = \frac{\tilde{m}_2(\tilde{y}_2 - \tilde{y}_1)}{\tilde{r}_{12}^3} + \frac{\tilde{m}_3(\tilde{y}_3 - \tilde{y}_1)}{\tilde{r}_{31}^3}, \quad \text{etc.,} \tag{6.49a,b}$$

where a dot denotes differentiation with respect to $\tilde{t}$.

We are free to choose M and d as we find convenient in any particular problem, but in what follows we shall usually choose $M = m_1 + m_2$ and d as the *initial* distance between m_1 and m_2, both of which are in keeping with the way in which we used M and d in Section 6.7.

The program THREEBP

The final step involves converting the six coupled second-order equations (6.49) into *twelve* coupled first-order equations, in the normal way. The program THREEBP, which is a modified version of the key program NPHASE, then integrates these twelve first-order equations by a double precision Runge–Kutta method.

As a test of the computation, THREEBP monitors and continually displays the dimensionless equivalent of the total energy (6.46), which is supposedly constant. Nonetheless, our usual check of re-running any integration with a smaller time step h over the same time interval remains as important as ever.

Errors introduced by the step-by-step method can build up spectacularly during 'close encounters', when one of the quantities $\tilde{r}_{12}$, $\tilde{r}_{23}$, $\tilde{r}_{31}$ becomes very small, and running THREEBP with a *fixed* time step h—even if h is quite small—typically ends in nonsense after a few such encounters.

The natural way round this problem is to modify THREEBP slightly so that the time step h automatically reduces itself whenever one or more of the quantities $\tilde{r}_{12}$, $\tilde{r}_{23}$, $\tilde{r}_{31}$ becomes small. We shall use the recipe

$$h = \frac{h_{\text{scale}}}{\tilde{r}_{12}^{-2} + \tilde{r}_{23}^{-2} + \tilde{r}_{31}^{-2}}, \tag{6.50}$$

where h_{scale} is fixed in advance. This is fairly arbitrary, but effective enough, apparently, for the purposes which follow.

Two examples

Suppose that the masses m_1 and m_2 are equal, and then choose $M = m_1 + m_2$ as our unit of mass, so that $\tilde{m}_1 = \tilde{m}_2 = 0.5$. Suppose too that m_1 and m_2 are initially on the x-axis, at equal distances from the origin. If we choose our unit of length d to be their initial distance apart, then we shall have $(\tilde{x}_1, \tilde{y}_1) = (-0.5, 0)$ and $(\tilde{x}_2, \tilde{y}_2) = (0.5, 0)$ at $\tilde{t} = 0$.

We begin by taking m_3 to be so small and far away from m_1 and m_2 that it has essentially no effect on them. According to (6.42) we may then make the equal masses m_1 and m_2 move around the origin in a circle if they have initial velocities of 0.5 $(GM/d)^{1/2}$. As our unit of velocity is $(GM/d)^{1/2}$ this means that the initial dimensionless velocities $(0, -0.5)$, $(0, 0.5)$ will result in a common circular orbit about the origin.

If we reduce these initial velocities from 0.5 to 0.3, m_1 and m_2 move in elliptical orbits with their centre of mass, which remains at the origin, as one focus (Fig. 6.15(a)).

Now take $m_3 = 0.5$, so that all three masses are equal, and let m_3 start at $(-0.1, 0.75)$ with velocity $(0, -0.3)$. We see from Fig. 6.15(b) that by $\tilde{t} = 0.85$ the mass m_3 has disrupted the previous motion, and has become involved in a close encounter with m_1. They subsequently experience another at $\tilde{t} = 1.85$, while m_2 is temporarily expelled. The masses m_1 and m_3 then revolve around one another until m_2 returns and experiences a close encounter with m_3 at $\tilde{t} = 4.85$. During each such close encounter the masses involved speed up and swing each other around with some ferocity; this can be quite striking when we run THREEBP in (roughly) 'real time', i.e. with a constant time-step h.

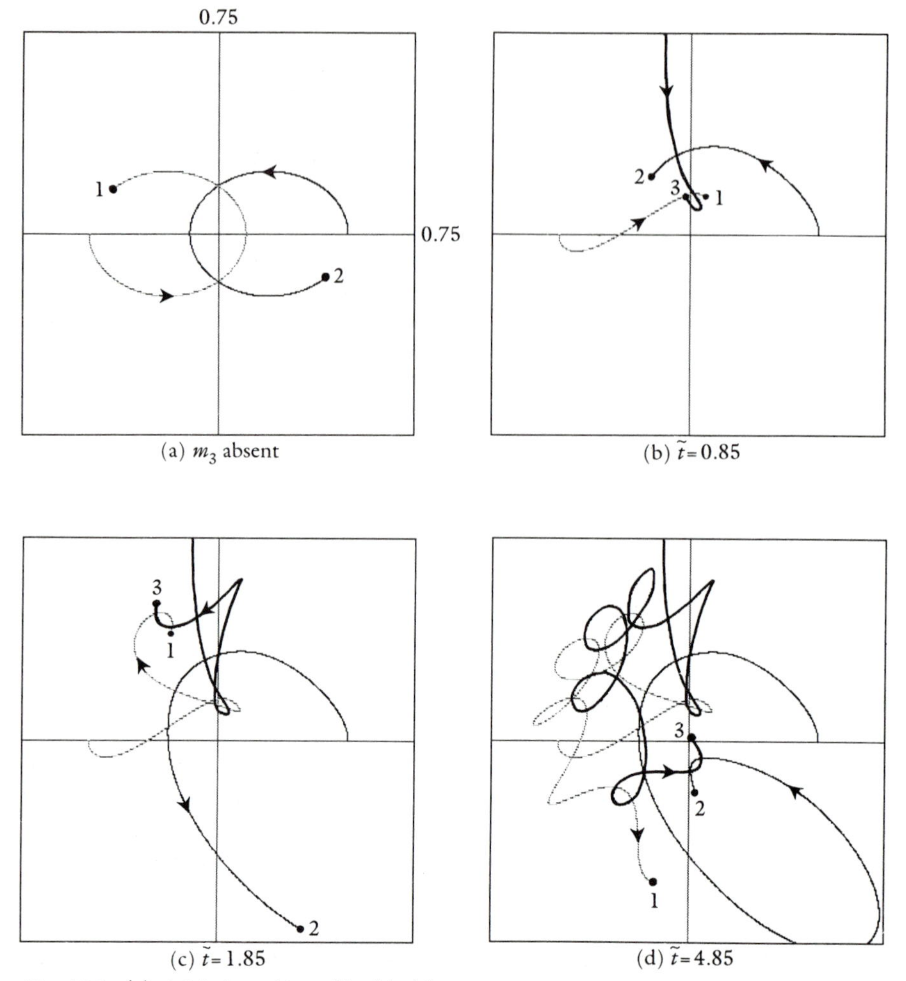

Fig. 6.15 (a) A 2-body problem. (b), (c), (d) A 3-body problem with initial positions $(-0.5, 0)$, $(0.5, 0)$, $(-0.1, 0.75)$ and initial velocities $(0, -0.3)$, $(0, 0.3)$, $(0, -0.3)$, all in dimensionless units ($\tilde{E}=0.737$).

Longer-term pairings can certainly arise, as we can see if we change the problem slightly, so that m_3 has an initial velocity of $(0, -0.2)$ rather than $(0, -0.3)$. A 3/1 pair again emerges, but it is more cohesive than before and m_2 looks at first as if it is going to leave the system. But, at $\tilde{t} \approx 20$, m_2 turns back and has a *very* close encounter with m_1 at $\tilde{t} = 23.56$, the distance between them falling to 0.00065 at closest approach. As a result of this, m_1 is expelled forcefully 'downwards', and a 2/3 pair heads up towards the origin. Both this pair and m_1 eventually turn back, leading to another close encounter at $\tilde{t} = 36.16$, as a result of which m_3 is expelled roughly in the negative x-direction, and a 1/2 pair is formed.

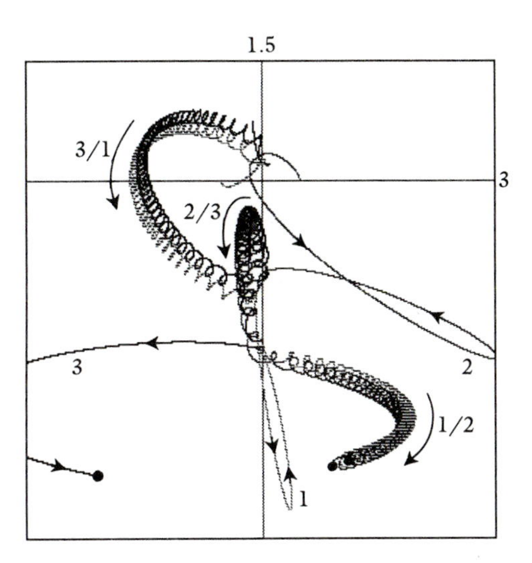

Fig. 6.16 Three successive longer-term 'pairings', shown on a larger scale. The initial conditions are as in Fig. 6.15, except that the initial velocity of m_3 is now (0, − 0.2).

These are only examples, of course, and the chief value of a program like THREEBP is that it can be used, with very little modification, to explore other dynamical problems of the same general type.

The restricted three-body problem

We end this chapter with a classical problem in which the mass m_3 is taken to be so small in comparison with m_1 and m_2 that its effect on their motion can be neglected. They therefore move according to two-body theory, and the problem that remains is to find the motion of m_3 under their gravitational influence. A good practical example of this is when m_3 corresponds to a 'spacecraft'.

For a brief look at this problem it is probably simplest to set $m_3 = 0$ in the program THREEBP. This does *not* lead to just a two-body problem, because the common factors of m_3 occurring on both sides of the last two equations of the set (6.44) have been cancelled out in advance (see (6.49)). In this way, setting $m_3 = 0$ in the program still gives twelve equations of first order, and in fact represents taking the *limit* $m_3 \to 0$, which is what we actually want.

We shall start the masses m_1 and m_2 on the $\tilde{x}$-axis at $(-\tilde{m}_2, 0)$ and $(\tilde{m}_1, 0)$, so that their centre of mass is at the origin and they are a unit distance apart. We also give these masses the (dimensionless) velocities $(0, -\tilde{m}_2)$ and $(0, \tilde{m}_1)$ which make them move in concentric *circles* around the origin (see (6.42)).

One long-standing question of interest is what kind of periodic or quasi-periodic motions of m_3 are then possible, and one example is shown in Fig. 6.17(a), with $\tilde{m}_1 = \tilde{m}_2 = 0.5$. The two 'primaries' move in the *same* circle, radius 0.5, in this case. Particle 3 starts at $(-1, 0)$ with initial velocity $(0, 0.506)$, and proceeds to orbit clockwise about m_1 three times during each

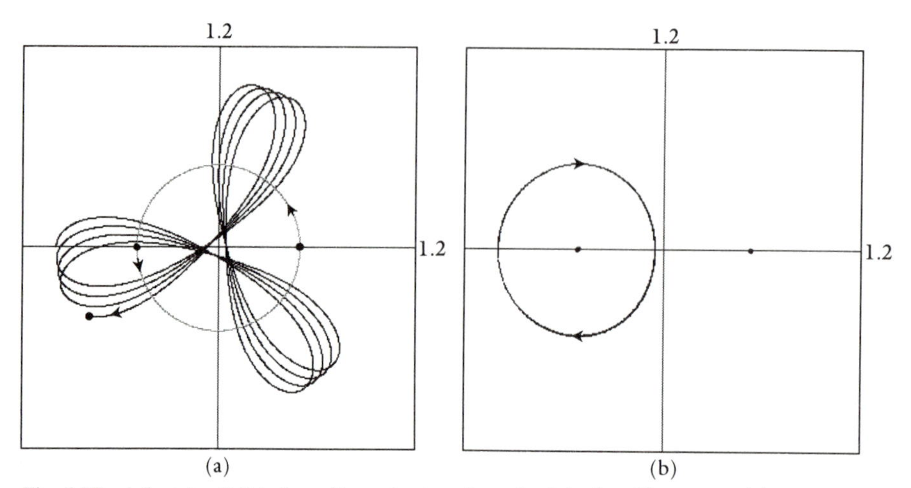

Fig. 6.17 A 'restricted' 3-body problem, showing the path of the (small) mass m_3 (a) in space and (b) relative to a frame rotating with the primaries; $\tilde{m}_1 = \tilde{m}_2 = 0.5$.

revolution of m_1 about the origin. Figure 6.17(b) shows the same motion of m_3 viewed from a frame of reference which rotates about the origin with the primaries, so that they now appear fixed. With respect to this frame, the motion of m_3 is clearly periodic.

More generally, m_3 can shuttle to and fro between the two revolving primaries, and Fig. 6.18 shows an example of this with $\tilde{m}_1 = 81/82$ and $\tilde{m}_2 = 1/82$, which corresponds roughly to the Earth–Moon system. The dominant mass m_1 hardly moves, while m_2 moves on a circle of almost unit radius. The 'spacecraft' m_3 is initially at $P = (1.05, 0)$ in Fig. 6.18(a), so that it

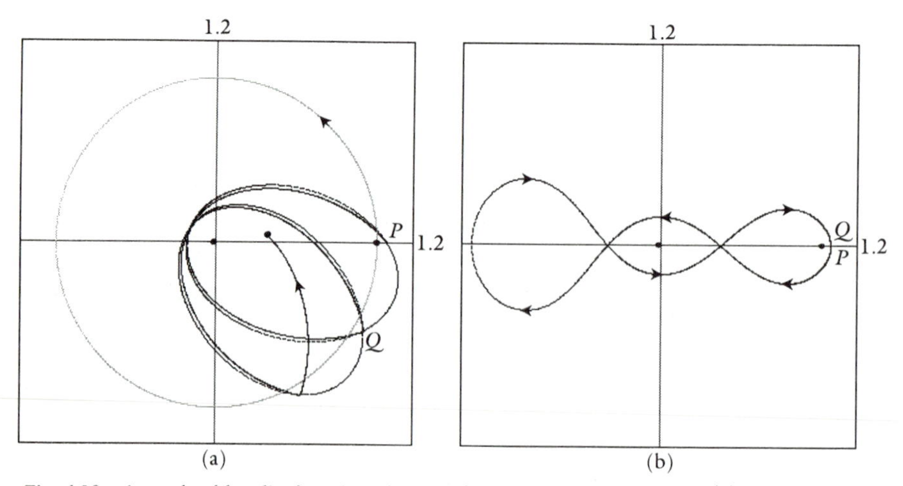

Fig. 6.18 A regular 'shuttling' motion of m_3 in the restricted 3-body problem (a) in space and (b) relative to a rotating frame; $\tilde{m}_1 = 81/82$, $\tilde{m}_2 = 1/82$.

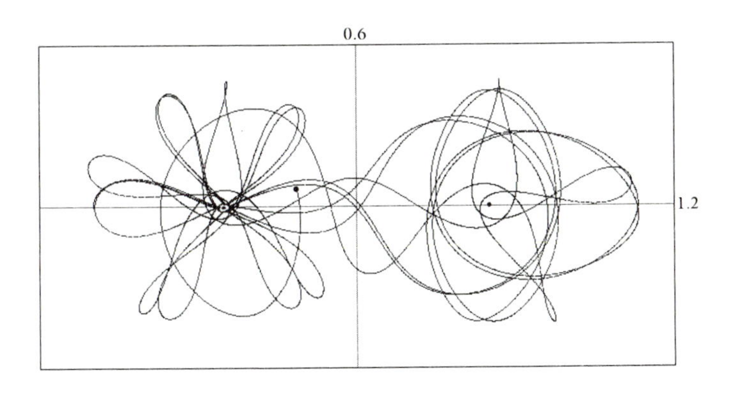

Fig. 6.19 Chaotic shuttling of m_3 in the restricted 3-body problem, relative to the rotating frame. The conditions are as for Fig. 6.17b, with m_3 again starting at $(-1, 0)$, except that its initial velocity is now $(0, -0.59)$ relative to the fixed frame, i.e. $(0, 0.41)$ relative to the rotating one.

starts close to m_2. Its initial velocity is $(0, 0.2012)$, but it quickly gets deflected by m_2, and as m_2 goes on its circular orbit m_3 loops around m_1 in an almost elliptical orbit. It then loops around m_1 *again* in a very slightly different orbit before meeting up with m_2 again at Q, while m_2 itself is still completing its first revolution about O. As a result of this new close encounter between m_3 and m_2 at Q, m_3 is severely deflected in just such a way that the sequence repeats. Figure 6.18(b) shows the same motion from the point of view of the rotating, or Earth–Moon, frame of reference.

More generally still in the restricted three-body problem, the small mass m_3 can shuttle to and fro between the two revolving primaries in an apparently haphazard or *chaotic* way, and Fig. 6.19 shows one example, with $m_1 = m_2$. Remarkable as it may seem, Poincaré deduced the existence of this kind of motion in the nineteenth century, long before the advent of electronic computers (see Chapter 11).

Exercises

6.1 A particle of mass m moves under an inverse-*cube* central force $f(r) = c/r^3$, where c is a constant. Initially it is at $r = d$, $\theta = 0$, with a radial component of velocity v and a transverse component $v_c = (c/md^2)^{1/2}$.

Find the path of the particle by the methods of Sections 6.4 and 6.5, and show that it takes the form shown in Fig. 6.20 when $v < 0$.

What form does it take if $v > 0$?

6.2 *Deducing* $f(r)$ *from the orbit*. A particle moves under a central force $f(r)$. Show that

(i) if its path is the spiral $r = e^{-k\theta}$, where k is a constant, then $f(r) \propto 1/r^3$,
(ii) if its path is a circular arc terminating at $r = 0$ then $f(r) \propto 1/r^5$.

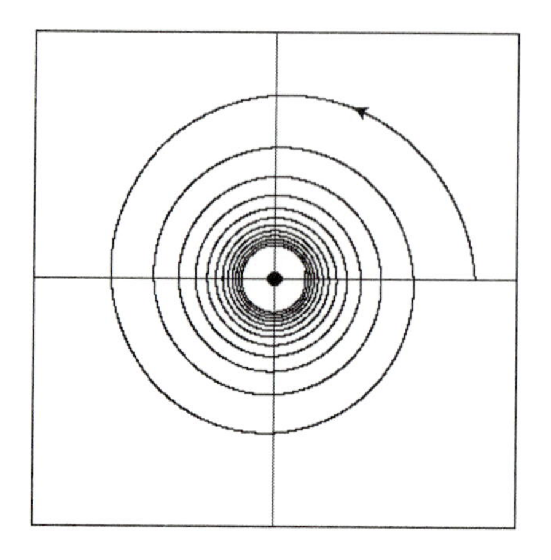

Fig. 6.20 A spiral path with $f(r) \propto r^{-3}$.

[Newton was much concerned with problems of this kind, and (ii) comes from Cor. 1 to Prop. VII, Problem II of Book I of his *Principia* (1687).]

6.3 $f(r) \propto 1/r^2$; *elliptical orbits*. Adapt the program NPHASE, as suggested on pp. 216–17, to confirm the results in Fig. 6.11. Then use a *variable time step* $h \propto r^2$ to solve the problem when $\tilde{v} = 0.3$, which involves a comparatively close approach to the origin.

Show that the dimensionless period of each elliptical orbit is

$$\tilde{T} = \frac{2\pi}{(2 - \tilde{v}^2)^{3/2}},$$

and use the program to confirm this for a selection of different values of $\tilde{v} = v/v_c$.

6.4 *A two-body problem*. Show that if two gravitating masses m_1, m_2 revolve in circular orbits about their centre of mass C with a common angular velocity Ω, as in Fig. 6.12, then

$$\Omega = \left(\frac{GM}{d^3}\right)^{1/2},$$

where $M = m_1 + m_2$ and d is their distance apart.

6.5 *Some three-body problems*. Use the program THREEBP to confirm the sequence in Fig. 6.15(b)–(d), and determine what happens next.

Now let $\tilde{m}_3$ be 0.1 instead, but with the same initial position and velocity. Compute the corresponding motion, and then investigate the effect of changing the initial position of m_3 by just one part in a thousand, to $(-0.1, 0.75075)$.

7 Waves and diffusion

7.1 Introduction

The aim of this chapter is to introduce some examples of **partial differential equations**, for these provide the key to some of the deepest phenomena in Nature.

Suppose, for instance, that we take a taut length of string, or skipping-rope, and draw a portion of it to one side. If we then release it, the initial displacement quickly splits into two separate **waves** which travel along the string in opposite directions (Fig 7.1).

Now, this kind of problem is clearly of a rather different kind to those we have discussed so far, because the displacement z depends on both time t *and the coordinate* x. There are, in other words, *two* independent variables here, x

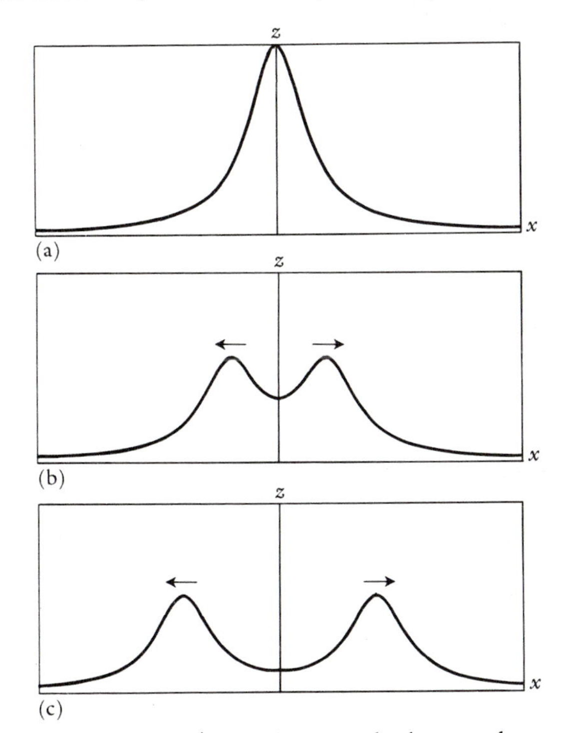

Fig. 7.1 Waves on a stretched string. (The displacement z has been greatly exaggerated.)

and t, and the way forward in such problems is to introduce the idea of

$$\textbf{partial derivatives:} \quad \frac{\partial z}{\partial x}, \quad \frac{\partial z}{\partial t}. \tag{7.1}$$

By each of these we simply mean the derivative of z with respect to *one* of the independent variables x and t *while holding the other one constant*. Thus if z were $x \sin \omega t$, for example, with ω a given constant, then $\partial z/\partial x$ would be $\sin \omega t$ and $\partial z/\partial t$ would be $x\omega \cos \omega t$.

Both the partial derivatives (7.1) have simple physical interpretations in our problem. The first, $\partial z/\partial x$, is the *slope* of the string at the position x and time t in question. In a similar way the second, $\partial z/\partial t$, represents the sideways *velocity* of the string.

It turns out, in fact, that small-amplitude motions of the string are governed by the equation

$$T\frac{\partial^2 z}{\partial x^2} = \rho \frac{\partial^2 z}{\partial t^2}, \tag{7.2}$$

where T denotes the tension in the string and ρ is its density, i.e. mass per unit length. The left-hand side represents the net force (in the z-direction) experienced by a small portion of the string, and the right-hand side is essentially mass $\times$ acceleration.

Now, *partial* differential equations like (7.2) occupy a central place in much of mathematical physics, and in order to study them at all systematically one must go beyond Chapter 2 to the **calculus of several variables.** Our aim here, however, is to motivate that kind of further study by taking a first look at partial differential equations using only elementary mathematical techniques.

7.2 Wave motion

Equation (7.2) is just one example of the so-called **wave equation**

$$\frac{\partial^2 z}{\partial t^2} = c^2 \frac{\partial^2 z}{\partial x^2}, \tag{7.3}$$

where c denotes a constant determined by the parameters of the system in question.

We begin our short study of (7.3) by examining one particularly simple solution.

An elementary solution

Consider the function

$$z = A \sin \frac{2\pi}{\lambda}(x - ct), \tag{7.4}$$

where A and λ are constants. On differentiating with respect to x while holding t constant we obtain

$$\frac{\partial z}{\partial x} = \frac{2\pi}{\lambda} A \cos \frac{2\pi}{\lambda}(x - ct),$$

and on doing so again we find

$$\frac{\partial^2 z}{\partial x^2} = -\frac{4\pi^2}{\lambda^2} A \sin \frac{2\pi}{\lambda}(x - ct).$$

If we then make a similar calculation of $\partial^2 z/\partial t^2$ we obtain almost the same results, but with an extra factor of $-c$ after one derivative and yet another factor of $-c$ after the second derivative. In this way we find that $\partial^2 z/\partial t^2$ is indeed equal to $c^2\, \partial^2 z/\partial x^2$, so (7.4) *is* a solution of (7.3).

Now consider what this solution actually looks like as time goes on. Initially, we have

$$z = A \sin \frac{2\pi}{\lambda} x \qquad \text{at } t = 0, \tag{7.5}$$

i.e. a sine curve (dotted in Fig. 7.2) with a distance λ between successive 'crests'. Now, according to (7.4), z is given at any later time t by the same expression *but with x replaced by $x - ct$*. This has the effect of shifting the whole curve by an amount ct in the positive x-direction, because x now has to be greater than before, by an amount ct, to produce any particular value of z. We now have, in other words, a *travelling wave*, and the speed with which the wave travels is c, the constant which occurs in the wave equation (7.3).

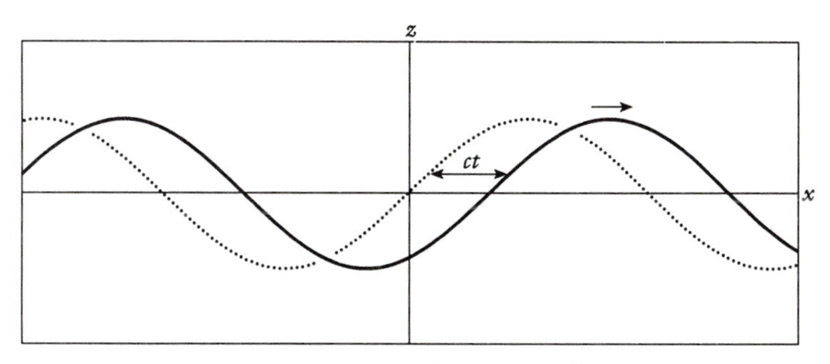

Fig. 7.2 The travelling wave (7.4).

Waves of general shape

The argument we have just given works equally well, in fact, for a waveform of *any* shape. To see this, consider

$$z = f(x - ct), \tag{7.6}$$

of which (7.4) is just one special case. This is, again, a wave travelling in the positive x-direction with speed c, because just as replacing $y=F(x)$ by $y=F(x-1)$ shifts that whole curve to the *right* by an amount 1, so (7.6) shifts the whole curve $z=f(x)$ to the right by an amount ct.

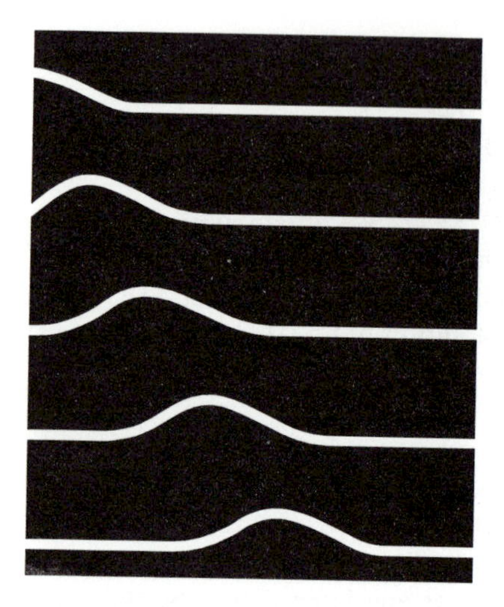

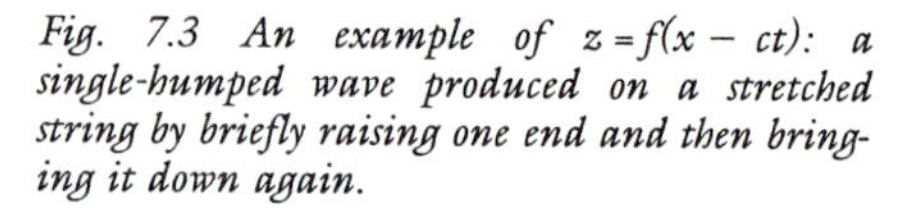

Fig. 7.3 An example of $z=f(x-ct)$: a single-humped wave produced on a stretched string by briefly raising one end and then bringing it down again.

To demonstrate that (7.6) satisfies the wave equation (7.3), write $X=x-ct$, so that

$$\frac{\partial z}{\partial x}=f'(X)\frac{\partial X}{\partial x}=f'(X)$$

and

$$\frac{\partial z}{\partial t}=f'(X)\frac{\partial X}{\partial t}=-cf'(X).$$

On proceeding to the second derivatives we find

$$\frac{\partial^2 z}{\partial x^2}=f''(X), \qquad \frac{\partial^2 z}{\partial t^2}=c^2 f''(X),$$

so

$$\frac{\partial^2 z}{\partial t^2}=c^2\frac{\partial^2 z}{\partial x^2}, \tag{7.7}$$

and (7.6) is therefore a solution of the wave equation (7.3) regardless of the *shape* of the wave, i.e. regardless of the form of the function $f(X)$.

In a similar way, any function of $x+ct$ is also a solution, and represents a wave travelling in the *negative* x-direction, again with speed c.

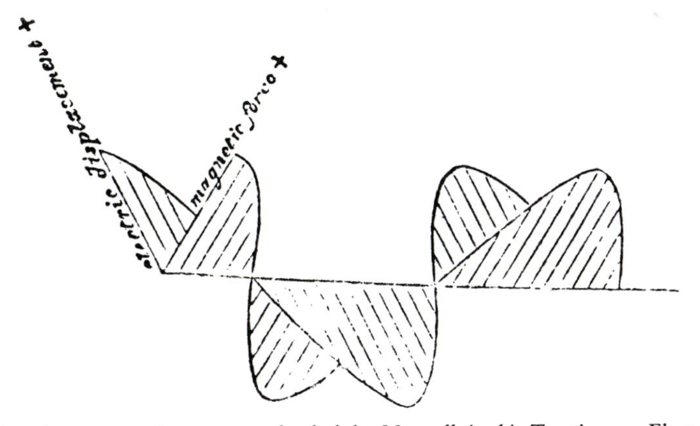

Fig. 7.4 An electromagnetic wave, as sketched by Maxwell in his Treatise on Electricity and Magnetism, *1873.*

Two applications

The first appearance of (7.3) was in a paper on vibrating strings by D'Alembert in 1745. In that particular case

$$c = \sqrt{\frac{T}{\rho}} \tag{7.8}$$

(see (7.2)), so the greater the tension in the string the faster the waves travel, as one would expect.

The wave equation subsequently emerged again in several other contexts, and a most notable reappearance was in connection with Maxwell's theory of electromagnetism. In around 1860 he showed that both the electric field and the magnetic field in free space satisfy the wave equation (7.3) with, on this occasion,

$$c = \frac{1}{\sqrt{\mu_0 \varepsilon_0}}. \tag{7.9}$$

The constants ε_0 and μ_0 could be determined in the laboratory; $1/\varepsilon_0$ is essentially a measure of how strong an electric field is produced by a given electric charge, and μ_0 is a measure of how strong a magnetic field is produced by a given electric current. In this way Maxwell calculated the electromagnetic wave speed (7.9) to be 193,088 miles per second. At the time, the best available direct measurement of the velocity of *light* was 193,118 miles per second, and these figures were so close that Maxwell inferred that light itself must be an electromagnetic phenomenon.

An initial-value problem

We will end this brief look at waves by returning to the original problem illustrated in Fig. 7.1.

Suppose, for instance, that the initial shape of the string there is

$$z = A\,\mathrm{e}^{-x^2/a^2} \qquad \text{at } t = 0, \tag{7.10}$$

which has a single hump centred on $x = 0$, as illustrated. Recall, too, that the string is initially at rest.

Now, we might perhaps conjecture

$$z = A\,\mathrm{e}^{-(x-ct)^2/a^2}$$

as the resulting motion; this would certainly be a solution of the wave equation—because it is of the form (7.6)—and it would satisfy the initial condition (7.10). But we would then have

$$\begin{aligned}\frac{\partial z}{\partial t} &= \frac{2A(x-ct)c}{a^2}\,\mathrm{e}^{-(x-ct)^2/a^2} \\ &= \frac{2Ax}{a^2}\,c\,\mathrm{e}^{-x^2/a^2} \qquad \text{at } t = 0,\end{aligned}$$

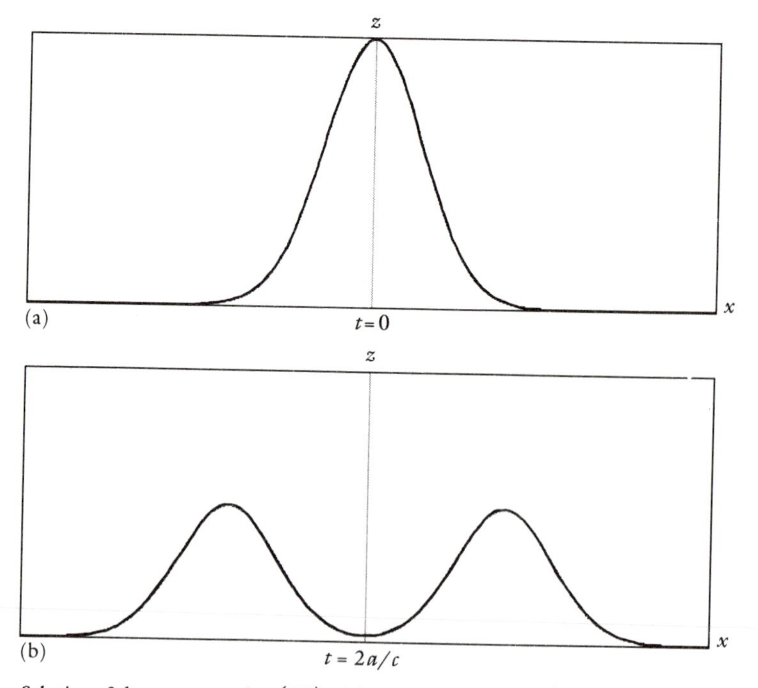

Fig. 7.5 Solution of the wave equation (7.3) with initial displacement (7.10) and $\partial z/\partial t = 0$ at $t = 0$.

so the initial velocity of the string would not be zero. Another possibility,

$$z = A\,\mathrm{e}^{-(x+ct)^2/a^2}$$

leads to a similar difficulty, i.e.

$$\frac{\partial z}{\partial t} = -\frac{2Ax}{a^2}c\,\mathrm{e}^{-x^2/a^2} \qquad \text{at } t = 0.$$

At this point, however, we notice that by combining *both* these waves (or, rather, half of each) in the following way

$$z = \tfrac{1}{2}A\,\mathrm{e}^{-(x-ct)^2/a^2} + \tfrac{1}{2}A\,\mathrm{e}^{-(x+ct)^2/a^2} \tag{7.11}$$

we solve the problem, for the two contributions to the initial velocity then cancel, and the string is initially at rest, with the correct shape.

In this way, then, we can understand the 'splitting' of the initial disturbance in Fig. 7.1, for (7.11) is none other than a superposition of two single-humped waves, travelling at equal speed c but in opposite directions.

7.3 The diffusion equation

For the purposes of comparison, we now take a brief look at another famous partial differential equation of physics, the so-called **diffusion equation**

$$\frac{\partial T}{\partial t} = \kappa\frac{\partial^2 T}{\partial x^2}, \tag{7.12}$$

where κ is a positive constant.

At first sight, perhaps, this is not unlike (7.3), but the properties of this equation do in fact turn out to be completely different. It first arose in 1822 in connection with Fourier's study of the diffusion of *heat* through a solid, and in that context T denotes the temperature and κ is a measure of how good the solid is as a thermal conductor.

To see broadly what (7.12) is saying, note first that heat flows in a solid from hot regions to cold regions at a rate which is *proportional to the local temperature gradient*. The rate of heat flow in the positive x-direction is therefore proportional to $-\partial T/\partial x$. Suppose, then, that at some particular time $-\partial T/\partial x$ decreases with x, so that $\partial^2 T/\partial x^2 > 0$. Then heat is passing the cross-section at x at a greater rate than it is passing the cross-section at $x + \delta x$, so the solid in between must be getting hotter with time. This is just what (7.12) says: if $\partial^2 T/\partial x^2 > 0$ then $\partial T/\partial t > 0$.

Another initial-value problem

In order to emphasize the difference between the diffusion equation (7.12) and the wave equation (7.3), let us take an initial condition

$$T = T_0\,\mathrm{e}^{-x^2/a^2} \qquad \text{at } t = 0 \tag{7.13}$$

deliberately like (7.10), so that the initial state is a localized 'hot' region, centred on the origin $x = 0$. The solution to the diffusion equation (7.12) which satisfies this initial condition turns out to be

$$T = \frac{T_0}{\left(1 + \dfrac{4\kappa t}{a^2}\right)^{1/2}} e^{-x^2/(a^2 + 4\kappa t)} \tag{7.14}$$

(see Ex. 7.3) and is sketched in Fig. 7.6 at two different times t.

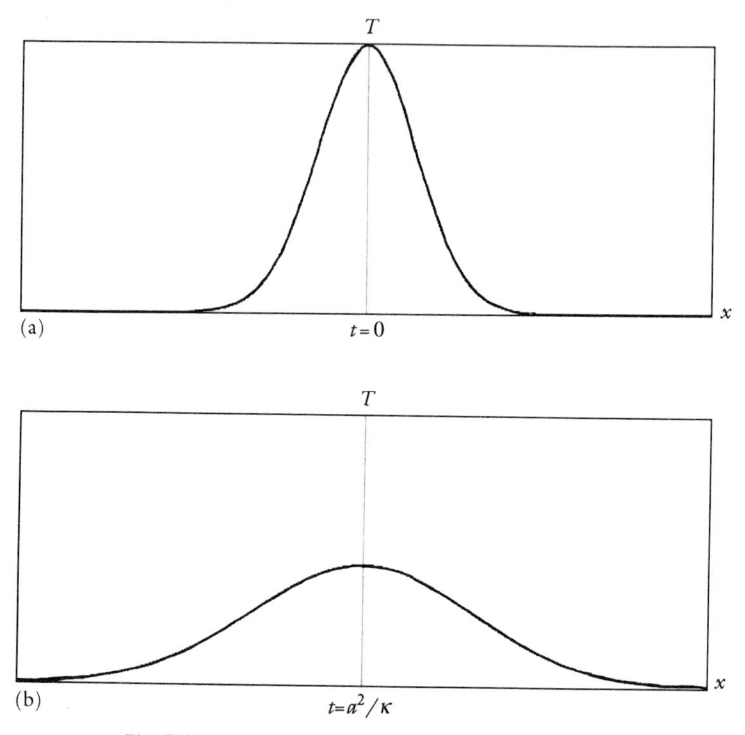

Fig. 7.6 Diffusion of heat, with initial condition (7.13).

The solution clearly evolves in a quite different way from that shown in Fig. 7.5. There is no wave-like behaviour, and the heat simply *spreads out* on a time scale of order a^2/κ, which will be small if κ is large, i.e. if the solid is a good conductor of heat. This is, of course, precisely what we would expect from a physical, rather than mathematical, point of view.

A numerical approach

Some of the computational solution methods of Chapter 4 can certainly be extended to *partial* differential equations, and we now illustrate this by

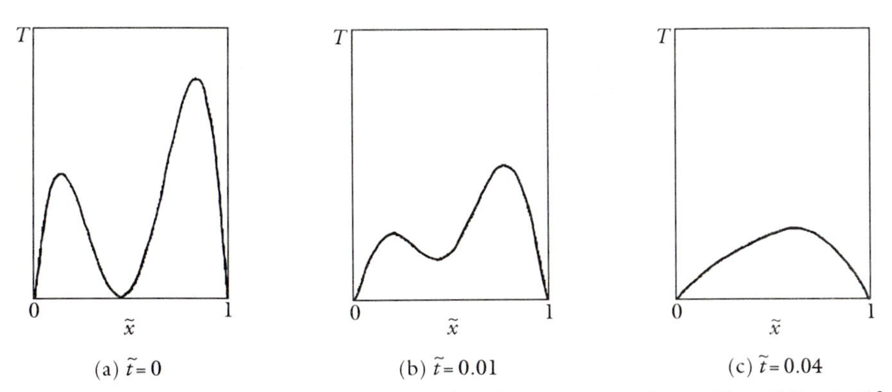

Fig. 7.7 Numerical solution of the heat equation (7.16) in $0<\tilde{x}<1$ with $T=40\tilde{x}(1-\tilde{x})(\tilde{x}-0.45)^2$ at $\tilde{t}=0$.

considering the diffusion equation (7.12) on the finite interval $0<x<l$. We will suppose that T is zero at the two end-points, $x=0,l$.

It is convenient first to introduce the dimensionless variables

$$\tilde{x}=\frac{x}{l}, \qquad \tilde{t}=\frac{\kappa t}{l^2}, \tag{7.15}$$

so that (7.12) becomes

$$\frac{\partial T}{\partial \tilde{t}}=\frac{\partial^2 T}{\partial \tilde{x}^2}, \tag{7.16}$$

with

$$T=0 \qquad \text{at } \tilde{x}=0,1 \tag{7.17}$$

and, say,

$$T=F(\tilde{x}) \qquad \text{at } \tilde{t}=0. \tag{7.18}$$

The program HEAT in Appendix B tackles this problem in the following way.

First we set up $\tilde{x},\tilde{t}$ coordinates and seek an approximation $T_{i,j}$ to the value of T at each point $(\tilde{x}_i,\tilde{t}_j)$ of the rectangular grid shown in Fig. 7.8(a). Thus the interval $0\le\tilde{x}\le 1$ is divided into m parts of length $h=1/m$, with

$$\tilde{x}_i=ih, \qquad i=0,1,\ldots,m, \tag{7.19}$$

and each time step is of amount k, so that

$$\tilde{t}_j=jk, \qquad j=0,1,2\ldots. \tag{7.20}$$

Next, we approximate the partial derivatives as follows:

$$\begin{aligned}\frac{\partial T}{\partial \tilde{t}} &\doteqdot \frac{T_{i,j+1}-T_{i,j}}{k},\\ \frac{\partial T}{\partial \tilde{x}} &\doteqdot \frac{T_{i+1,j}-T_{i,j}}{h},\end{aligned} \tag{7.21a,b}$$

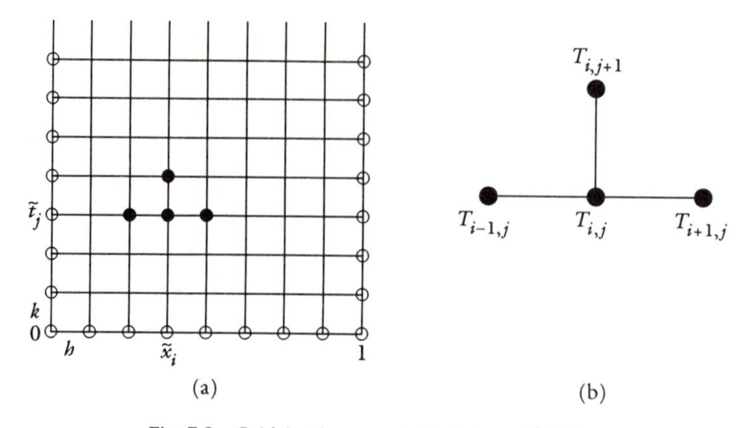

Fig. 7.8 Grid for the numerical solution of (7.16).

(see Fig. 7.8(b)), where i runs from 1 to $m-1$. These approximations are quite crude, and akin to those of the Euler method in Section 4.2. Finally, we approximate the *second* derivative on the right-hand side of (7.16) in the same kind of way, by subtracting two neighbouring approximations to $\partial T/\partial \tilde{x}$ and dividing by the distance h between the corresponding grid points. This gives

$$\frac{T_{i,j+1}-T_{i,j}}{k}=\frac{1}{h}\left[\frac{T_{i+1,j}-T_{i,j}}{h}-\left(\frac{T_{i,j}-T_{i-1,j}}{h}\right)\right] \tag{7.22}$$

as our finite-difference representation of the heat equation (7.16), with i again running from 1 to $m-1$. In a slightly different notation (7.22) gives

$$T_i^{\text{new}}=T_i^{\text{old}}+\frac{k}{h^2}(T_{i+1}^{\text{old}}-2T_i^{\text{old}}+T_{i-1}^{\text{old}}), \qquad i=1,2,\dots m-1 \tag{7.23}$$

as our 'updating' algorithm, each new value of T_i being calculated from the *three* nearest old values of T_i (Fig. 7.8(b)). At each encircled grid point in Fig. 7.8(a) we know $T_{i,j}$, whether from (7.17) or (7.18), and (7.22) then allows us to advance the solution, in principle, over and over again. Some typical results are shown in Fig. 7.7, where two 'hotspots' diffuse into one another and decay.

There are now, of course, *two* step sizes h and k which we must reduce when looking for a degree of convergence, and we need to be rather careful how we do this, for it turns out that the updating algorithm (7.23) is only *stable* if

$$\frac{k}{h^2}\leq 0.5. \tag{7.24}$$

If this condition is not satisfied, wild and completely spurious oscillations develop, and this is just one small indication of how partial differential equations can be more tricky to solve numerically than the ordinary differential equations of, say, Chapter 4.

Two-dimensional diffusion

It may be, of course, that the temperature T within a solid depends on *two* space variables x and y, say, as well as time t. In that case $T(x, y, t)$ satisfies

$$\frac{\partial T}{\partial t} = \kappa \left(\frac{\partial^2 T}{\partial x^2} + \frac{\partial^2 T}{\partial y^2} \right). \tag{7.25}$$

This two-dimensional diffusion equation can, again, be solved by numerical methods, and in Fig. 7.9 we display the solution in the square region $0 < x < l$,

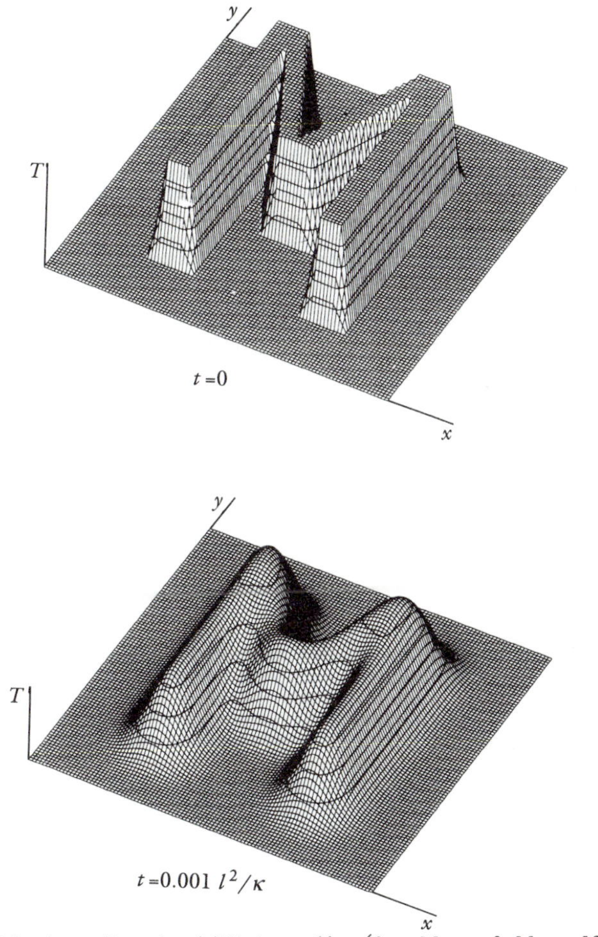

Fig. 7.9 A two-dimensional diffusion problem (from Morton & Mayers 1994).

$0 < y < l$, with $T = 0$ on the boundary of the square and an initially hot region in the shape of the letter M. The initial temperature gradients are steeper than those in Fig. 7.7, so the diffusion proceeds faster at first, and major changes are evident even when $t = 0.001l^2/\kappa$.

How the leopard gets its spots?

Surprisingly, perhaps, diffusion processes can sometimes help *create* patterns, rather than just destroy them.

According to a recent model by Jim Murray of the University of Washington, some animal coat markings can be explained by assuming the presence of just two chemical 'morphogens' which react and diffuse. Their concentrations u and v are then governed by a pair of equations of the form

$$\begin{aligned} \frac{\partial u}{\partial t} &= f(u,v) + \kappa_1\left(\frac{\partial^2 u}{\partial x^2} + \frac{\partial^2 u}{\partial y^2}\right), \\ \frac{\partial v}{\partial t} &= g(u,v) + \kappa_2\left(\frac{\partial^2 v}{\partial x^2} + \frac{\partial^2 v}{\partial y^2}\right). \end{aligned} \tag{7.26}$$

Each of these is a bit like (7.25), but they are coupled through the terms $f(u,v)$ and $g(u,v)$, which are nonlinear functions of u and v, their precise form depending on what one assumes about how the two morphogens react.

Remarkably, spatial patterns of the morphogen concentrations can emerge spontaneously from these equations by a mechanism first proposed by Alan Turing in 1952. Some typical patterns are shown in Fig. 7.10; other things being equal, the final 'spottiness' is greater for a large animal than for a smaller one of the same shape.

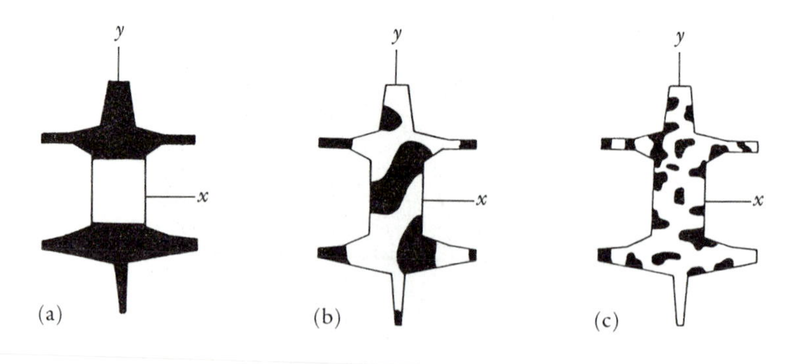

Fig. 7.10 How the leopard gets its spots? The three 'animal' sizes are in roughly the ratio 1:3:7. (After Murray 1989).

A most intriguing feature of this mechanism is that it is a *doubly*-diffusive one, and only works if the two morphogens diffuse at different rates, i.e. if $\kappa_1 \neq \kappa_2$ in (7.26).

Exercises

7.1 *Partial derivatives*. Calculate $\partial z/\partial x$ and $\partial z/\partial t$ in the three cases

(i) $z = x^2$
(ii) $z = (x - ct)^2$
(iii) $z = \dfrac{1}{1 + (x - ct)^2}$,

where c is a constant.

7.2 *Standing waves*. Suppose that a stretched string is fixed at two end-points $x = 0$ and $x = l$, say, so that $z = 0$ there for all time t. Investigate the natural modes of vibration of the string by seeking solutions to the wave equation (7.2) of the form

$$z = f(x)\sin \omega t$$

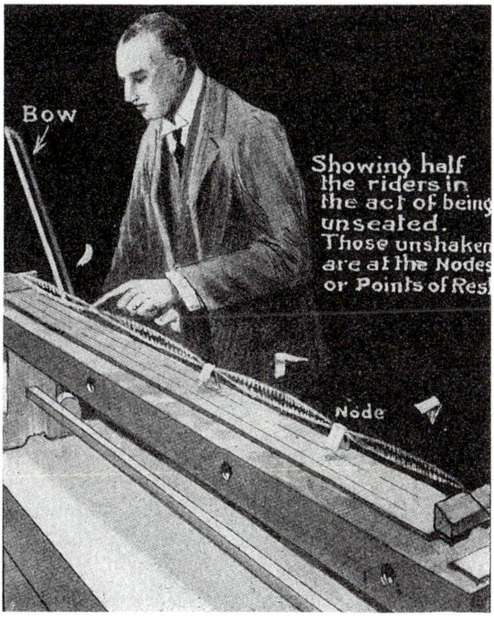

Fig. 7.11 Standing waves on a stretched string. (From The World of Sound *by W. H. Bragg, 1920.)*

and deducing that

$$T\frac{\mathrm{d}^2 f}{\mathrm{d}x^2} + \rho\omega^2 f = 0.$$

Solve this equation, subject to the boundary conditions at $x = 0$ and l, and show that such vibrations can occur only at certain *natural frequencies*

$$\frac{\omega}{2\pi} = \frac{N}{2l}\left(\frac{T}{\rho}\right)^{1/2}, \qquad N = 1, 2, 3\ldots \tag{7.27}$$

Show too that the higher the value of N the larger the number of *nodes*, i.e. points of zero displacement on the string—a result well known to musicians.

7.3 *Diffusion*. (i) Explain why (7.14) implies that the heat *spreads out* as time goes on, as indicated in Fig. 7.6.

(ii) Verify that

$$T = \frac{1}{t^{1/2}}\,\mathrm{e}^{-x^2/4\kappa t}$$

is a solution of the diffusion equation (7.12). [That (7.14) is a solution can be established in a similar way.]

7.4 Use the program `HEAT` to confirm the results in Fig. 7.7 and to explore the effect of different initial conditions.

7.5 *The calculus of several variables*. In order to take partial differential equations—and indeed several other subjects in this book—any further, we need to study the calculus of *functions of several variables*.

To give some hint of how this goes, suppose $\phi(x, y)$ is a function of two variables x and y, and suppose that $x(t)$ and $y(t)$ are both functions of time t. Then ϕ itself can be regarded as a function of t, and it turns out that

$$\frac{\mathrm{d}\phi}{\mathrm{d}t} = \frac{\partial\phi}{\partial x}\frac{\mathrm{d}x}{\mathrm{d}t} + \frac{\partial\phi}{\partial y}\frac{\mathrm{d}y}{\mathrm{d}t}. \tag{7.28}$$

Confirm this extended *chain rule*, which is crucial to the whole theory, in the particular case $\phi(x, y) = xy + y^2$, $x = t$, $y = t^2$.

8 The best of all possible worlds?

8.1 Introduction

Suppose that a ray of light is reflected by a mirror from some given point A to another given point B. It is well known that the point of reflection, P, is such that the rays AP and PB make equal angles with the mirror (Fig. 8.1). But there is in fact a quite different and more intriguing way of expressing the same physical result, namely that the point of reflection P is such that the total distance $AP + PB$ is *as small as possible*.

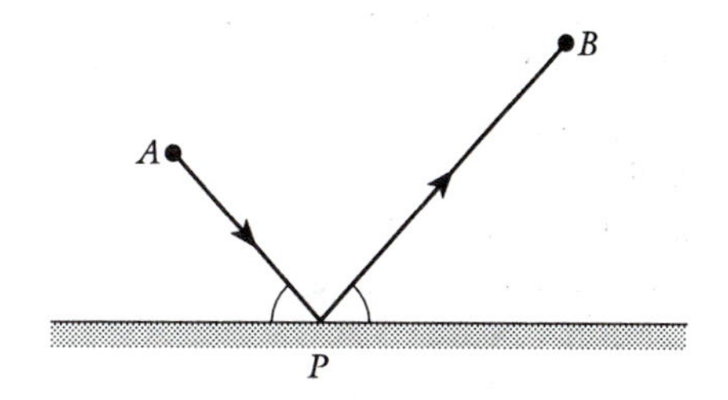

Fig. 8.1 Reflection at a plane mirror.

This was first discovered by Heron of Alexandria (c. AD 100), and his proof involves a simple but inspired geometrical construction (Fig. 8.2). If A' is the image point of A, so that $AP = A'P$, then the problem of choosing P to minimize $AP + PB$ is the same as the problem of minimizing $A'P + PB$. But

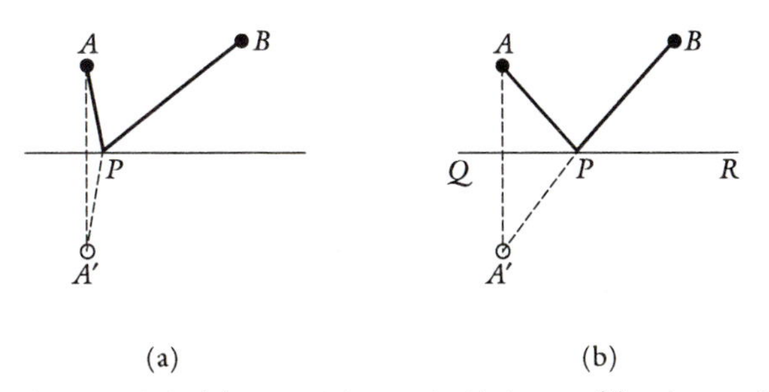

Fig. 8.2 Reflection of light (a) as it might conceivably happen (b) as it actually happens.

this is clearly solved by choosing P such that $A'PB$ is a straight line. Then $\angle BPR = \angle A'PQ$, and as $\angle A'PQ = \angle APQ$ we have $\angle BPR = \angle APQ$, so that the rays AP and PB make equal angles with the mirror.

Heron's result remained a more or less isolated curiosity until 1661, when Fermat found that both the reflection and the refraction of light appeared to obey a single principle of least time, rather than least distance (Ex. 8.1). This in turn eventually prompted the question of whether there might be some equivalent 'minimum principle' governing problems of *mechanics*, and in 1744 the French mathematician Maupertuis put forward a principle of *least action*.

The main driving force behind this idea was theological. If Nature were expending as little 'action' as possible, then it would be working, in some sense, as 'perfectly' as possible, and for Maupertuis this meant

... proof of the existence of Him who governs the world.

This philosophical idea is somewhat reminiscent of Leibniz's earlier contention that we might be living in 'the best of all possible worlds,' which had itself been met with some scepticism at the time. It was most famously attacked in Voltaire's novel *Candide* (1759), and in due course Voltaire poured scorn on Maupertuis's philosophy also, writing

We ask forgiveness of God for having pretended that there is only proof of his existence in A + B divided by Z, etc...

For present purposes, however, the main weakness of Maupertuis's least action principle is that he never really gives a proper definition of what action *is*, and he seems, almost, to adjust the concept to his convenience as he passes from one particular physical problem to another.

Before proceeding further, therefore, our first task must be to clarify just what is meant by the term action.

8.2 The concept of 'action'

The modern definition of the total **action** S during some change in a physical system is the time integral of the *difference* between the kinetic energy T and the potential energy V:

$$S = \int_{t_1}^{t_2} T - V \, \mathrm{d}t. \qquad (8.1)$$

An example of 'least action'

Suppose that a particle of mass m is at height y_1 at time t_1 and then moves vertically, under gravity, so that it is at height y_2 at time t_2. Then $T = \frac{1}{2}m\dot{y}^2$ and $V = mgy$, so

$$S = \int_{t_1}^{t_2} \left(\tfrac{1}{2}m\dot{y}^2 - mgy\right) \mathrm{d}t. \qquad (8.2)$$

We will now show that in this particular example the *actual* value of S is *smaller than it would have been if the motion $y(t)$ had taken place between the same two points, in the same time interval, in any other way.*

To this end, let $y_A(t)$ be the 'actual' motion, so that

$$\ddot{y}_A = -g, \qquad \text{with} \qquad \begin{aligned} y_A(t_1) &= y_1 \\ y_A(t_2) &= y_2, \end{aligned} \tag{8.3}$$

and denote any 'alternative' motion—which will be wholly fictitious and will *not* be required to satisfy any dynamical laws—by

$$y(t) = y_A(t) + \eta(t). \tag{8.4}$$

We stipulate only that in any of these fictitious alternative motions the particle should start and finish at the right place, at the right time, so that

$$\eta(t_1) = \eta(t_2) = 0 \tag{8.5}$$

(see Fig. 8.3).

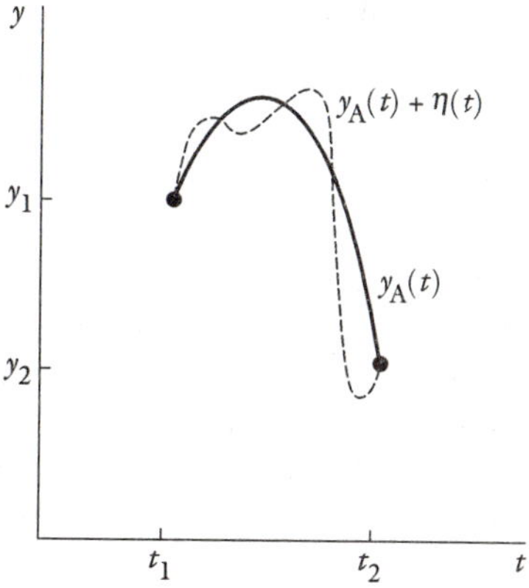

Fig. 8.3 Actual (——) and fictitious (or 'varied') motion (- - -) of a particle moving in a vertical line under gravity between two prescribed points in a prescribed time interval.

On substituting (8.4) into (8.2) we find

$$\begin{aligned} S &= m \int_{t_1}^{t_2} \left[\tfrac{1}{2}(\dot{y}_A + \dot{\eta})^2 - g(y_A + \eta) \right] \mathrm{d}t \\ &= S_A + m \int_{t_1}^{t_2} (\dot{y}_A \dot{\eta} - g\eta)\, \mathrm{d}t + \tfrac{1}{2}m \int_{t_1}^{t_2} \dot{\eta}^2 \, \mathrm{d}t, \end{aligned} \tag{8.6}$$

where S_A denotes the numerical value of S for the motion which actually occurs, obtained by putting $y = y_A(t)$ in (8.2).

Now, the second term on the right hand side of (8.6) lends itself to integration by parts:

$$m[\dot{y}_A \eta]_{t_1}^{t_2} - m \int_{t_1}^{t_2} (\ddot{y}_A + g)\eta \, \mathrm{d}t.$$

But the first of these terms vanishes, because $\eta(t_1) = \eta(t_2) = 0$, and the second vanishes because the motion that actually occurs satisfies $\ddot{y}_A = -g$ (see (8.3)). This leaves us with

$$S = S_A + \tfrac{1}{2}m \int_{t_1}^{t_2} \dot{\eta}^2 \, dt,$$

and because $\dot{\eta}^2 \geq 0$ throughout the interval $t_1 \leq t \leq t_2$ we find

$$S \geq S_A, \tag{8.7}$$

with equality only if $\dot{\eta}$ is zero throughout the time interval $t_1 \leq t \leq t_2$. Equality in (8.7) is only possible, then, if η is a constant during that time interval, and in view of the conditions (8.5) this is only possible if $\eta = 0$. This proves that S is least for the motion which is actually observed.

The principle of stationary action

Despite the above example, and many others like it, it is simply not always true that the action is least for the observed motion of a dynamical system. Writing on the matter in 1833, Hamilton noted that

> ... the quantity pretended to be economized is in fact often lavishly expended...,

for he knew of dynamical systems where S, as defined by (8.1), is *greatest* for the motion which actually occurs.

It turns out that we cannot in general be sure that S will be either a minimum *or* a maximum; we can claim only that the action S will be *stationary** with respect to small (fictitious) changes in the motion that actually occurs. This is the **principle of stationary action**, also known as **Hamilton's principle.**

If it lacks the philosophical or even theological overtones of Maupertuis's original idea, Hamilton's principle is, nonetheless, a powerful unifying concept in modern dynamics, and it is particularly effective when combined with one of Euler's cleverest inventions: the calculus of variations.

8.3 The calculus of variations

Some of the most interesting topics in applied mathematics lead, eventually, to a problem of the following kind.

We wish to find a function $y(x)$ which takes given values at the given 'end-points' $x = x_1$ and $x = x_2$, and which minimizes (or maximizes) the definite integral

$$I = \int_{x_1}^{x_2} F(x, y, \dot{y}) \, dx. \tag{8.8}$$

* This idea is similar to the simpler notion of a quantity $y = f(x)$ being stationary with respect to small changes in x (Ex. 2.1).

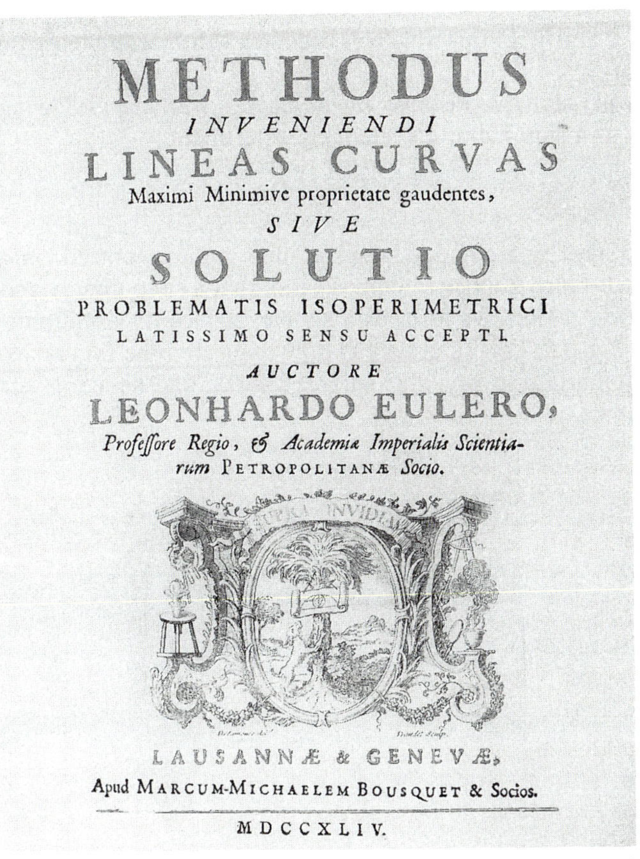
METHODUS
INVENIENDI
LINEAS CURVAS
Maximi Minimive proprietate gaudentes,
SIVE
SOLUTIO
PROBLEMATIS ISOPERIMETRICI
LATISSIMO SENSU ACCEPTI.
AUCTORE
LEONHARDO EULERO,
Professore Regio, & Academiæ Imperialis Scientiarum PETROPOLITANÆ Socio.
LAUSANNÆ & GENEVÆ,
Apud MARCUM-MICHAELEM BOUSQUET & Socios.
MDCCXLIV.

Fig. 8.4 The title page from Euler's 1744 treatise on the calculus of variations.

Here $F(x, y, \dot{y})$ is some *given* function of x, y and $\dot{y}$, where $\dot{y} = \mathrm{d}y/\mathrm{d}x$.

This is the central problem of the **calculus of variations**, which Euler described as the art of 'finding curved lines which enjoy some maximum or minimum property.' It is an altogether more difficult problem than the classical one of finding some *number* x so as to maximize or minimize some quantity $f(x)$. Nonetheless, we begin in a rather similar way, by seeking the equivalent of the condition for a stationary point, $f'(x) = 0$.

This turns out to be the **Euler–Lagrange equation**

$$\frac{\mathrm{d}}{\mathrm{d}x}\left(\frac{\partial F}{\partial \dot{y}}\right) - \frac{\partial F}{\partial y} = 0, \tag{8.9}$$

which is, despite appearances, an ordinary differential equation for the unknown function $y(x)$. To see this, recall that $F(x, y, \dot{y})$ in (8.8) is some *given* function of x, y and $\dot{y}$, so the partial derivatives $\partial F/\partial y$ and $\partial F/\partial \dot{y}$ can then be calculated and are therefore *known* functions of the same three variables x, y and $\dot{y}$. In this way (8.9) becomes a differential equation for y as a function of x.

All this will, perhaps, become much clearer with the following example, which leads to a simple but entertaining experiment.

Example: a soap film

Suppose we have two circular hoops of unit radius, centred on a common x-axis and a distance $2a$ apart. Suppose, too, that a soap film extends between the two hoops, taking the form of a simple surface of revolution about the x-axis, as in Fig. 8.5. Then if gravity may be neglected the film takes up a state of stable equilibrium in which its surface energy, and hence its surface area, is a minimum.

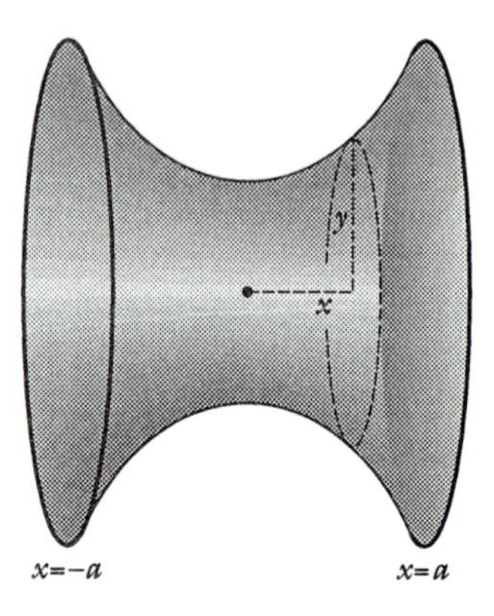

Fig. 8.5 A soap film spanning the gap between two circular hoops.

Our problem, then, is to find the function $y(x)$, satisfying the end conditions

$$y(-a)=y(a)=1, \qquad (8.10)$$

which makes the surface area

$$A=2\pi\int_{-a}^{a} y(1+\dot{y}^2)^{1/2}\,\mathrm{d}x \qquad (8.11)$$

a minimum. Thus

$$F(x,y,\dot{y})=2\pi y(1+\dot{y}^2)^{1/2}, \qquad (8.12)$$

and F does not in fact depend explicitly on x in this particular case.

Now, the partial derivative $\partial F/\partial y$ in (8.9) denotes the derivative of F with respect to y while holding the other two variables, namely x and $\dot{y}$, constant. So

$$\frac{\partial F}{\partial y} = 2\pi(1+\dot{y}^2)^{1/2}, \tag{8.13}$$

and in a similar way

$$\frac{\partial F}{\partial \dot{y}} = 2\pi y \cdot \tfrac{1}{2}(1+\dot{y}^2)^{-1/2} \cdot 2\dot{y} = \frac{2\pi y\dot{y}}{(1+\dot{y}^2)^{1/2}}. \tag{8.14}$$

The Euler–Lagrange equation (8.9) therefore becomes

$$\frac{\mathrm{d}}{\mathrm{d}x}\left\{\frac{y\dot{y}}{(1+\dot{y}^2)^{1/2}}\right\} - (1+\dot{y}^2)^{1/2} = 0, \tag{8.15}$$

and this is, as we claimed, an ordinary differential equation for $y(x)$.

It can in fact be put into the much simpler form

$$y\ddot{y} - \dot{y}^2 = 1, \tag{8.16}$$

which we must then solve subject to the boundary conditions (8.10), i.e.

$$y(-a) = y(a) = 1. \tag{8.17}$$

While (8.16) is second order and nonlinear, it is autonomous, and may therefore be tackled by the method in Section 3.5. The most interesting feature of the result is that the above problem *only has a real solution if*

$$a < 0.6627 \tag{8.18}$$

(Ex. 8.2).

Theory predicts, then, that as we gradually move the rings further apart, the soap film will suddenly collapse to some quite different state when the distance between the rings exceeds 0.6627 of their common diameter. This can be confirmed quite well even with a fairly casual 'kitchen' experiment.

To re-cap, the problem of finding the function $y(x)$ which makes the integral (8.8) stationary is tackled by solving the differential equation (8.9). To *prove* this requires techniques which are beyond the scope of this book, but we have at least seen in our soap-film example that all works out well.

In the next section we will use this connection between (8.8) and (8.9) to link Hamilton's principle of stationary action to a famous set of dynamical equations.

8.4 Lagrange's equations of motion

The following novel method for treating quite general dynamical systems stems, essentially, from Lagrange's *Mécanique Analytique* of 1788, and it uses *energy*, rather than force, as the fundamental concept.

Suppose, then, that we have a dynamical system with N degrees of freedom, so that N independent variables $q_1, \ldots, q_N$ are needed to specify the configuration of the system at any moment. We begin by obtaining general expressions for the kinetic energy T and the potential energy V of the system as a whole, and form the so-called *Lagrangian*

$$L = T - V, \tag{8.19}$$

which will then be some known function of the variables $q_i(t)$, their derivatives $\dot{q}_i(t)$, and possibly time t itself. Remarkable as it may seem, we may then write down the equations of motion without any further consideration of the mechanics or physics of the system. They are **Lagrange's equations**

$$\frac{\mathrm{d}}{\mathrm{d}t}\left(\frac{\partial L}{\partial \dot{q}_i}\right) - \frac{\partial L}{\partial q_i} = 0, \quad i = 1, \ldots, N, \tag{8.20}$$

i.e. N coupled ordinary differential equations for the variables $q_1(t), \ldots, q_n(t)$.

The most direct way of establishing (8.20) is by using Hamilton's principle (Section 8.2) and the calculus of variations; the equations (8.20) then follow from making (8.1) stationary, in much the same way that (8.9) can be shown to follow from making (8.8) stationary. But we will not dwell on this here; instead, we will illustrate the above method by two specific examples.

Fig. 8.6 J.-L. Lagrange (1736–1813).

A falling particle

Consider again the vertically moving particle of Section 8.2. This system has only one degree of freedom, and we need only one variable $q_1 = y$. The kinetic energy is $\frac{1}{2}m\dot{y}^2$ and the potential energy is mgy, so

$$L = \tfrac{1}{2}m\dot{y}^2 - mgy, \tag{8.21}$$

and (8.20) yields just one equation

$$\frac{\mathrm{d}}{\mathrm{d}t}\left(\frac{\partial L}{\partial \dot{y}}\right) - \frac{\partial L}{\partial y} = 0. \tag{8.22}$$

On calculating the partial derivatives we obtain

$$\frac{\partial L}{\partial \dot{y}} = m\dot{y}, \qquad \frac{\partial L}{\partial y} = -mg, \tag{8.23}$$

so (8.22) reduces to

$$m\ddot{y} = -mg, \tag{8.24}$$

which is indeed the equation of motion in this particular case.

Particle motion under a central force

As a second example of Lagrange's method, let us return to the problem of particle motion under a central force, first addressed in Section 6.2. Here there are *two* degrees of freedom, and $q_1 = r$ and $q_2 = \theta$ are our chosen

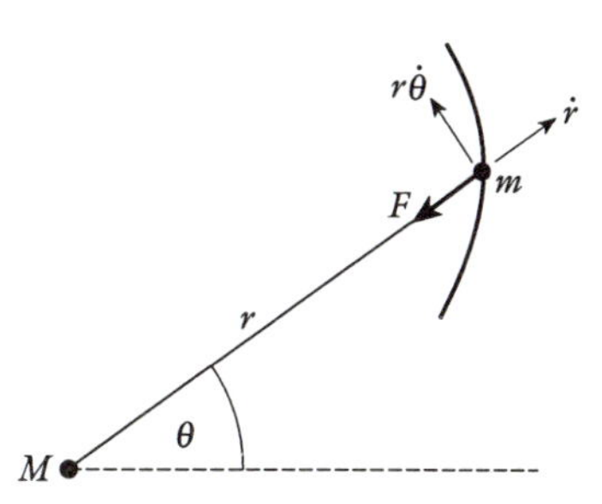

Fig. 8.7 Motion under a central force.

coordinates (see Fig. 8.7). In this case, then, (8.20) yields *two* Lagrange equations:

$$\begin{aligned} \frac{\mathrm{d}}{\mathrm{d}t}\left(\frac{\partial L}{\partial \dot{r}}\right) - \frac{\partial L}{\partial r} &= 0, \\ \frac{\mathrm{d}}{\mathrm{d}t}\left(\frac{\partial L}{\partial \dot{\theta}}\right) - \frac{\partial L}{\partial \theta} &= 0, \end{aligned} \tag{8.25a,b}$$

where L is to be viewed as a function of the four variables r, θ, $\dot{r}$, $\dot{\theta}$.

Now, the kinetic energy is $T = \frac{1}{2}m(\dot{r}^2 + r^2\dot{\theta}^2)$, and in the case of gravitational attraction towards a fixed mass M at the origin the potential energy is

$V = -GMm/r$ (see (6.39)). The Lagrangian for the system is therefore

$$L(r,\theta,\dot{r},\dot{\theta}) = \tfrac{1}{2}m(\dot{r}^2 + r^2\dot{\theta}^2) + \frac{GMm}{r}. \tag{8.26}$$

Next, we turn to (8.25a), and recall that $\partial L/\partial r$ means here the partial derivative with respect to r while holding θ, $\dot{r}$ and $\dot{\theta}$ constant. Thus

$$\frac{\partial L}{\partial r} = mr\dot{\theta}^2 - \frac{GMm}{r^2},$$

and in a similar way

$$\frac{\partial L}{\partial \dot{r}} = m\dot{r}.$$

So (8.25a) reduces to

$$m(\ddot{r} - r\dot{\theta}^2) = -\frac{GMm}{r^2}, \tag{8.27}$$

as we obtained by an entirely different argument in Chapter 6.

More interestingly still, we note that L—as given by (8.26)—*does not depend explicitly on* θ, so

$$\frac{\partial L}{\partial \theta} = 0, \tag{8.28}$$

as the partial derivative here means differentiation with respect to θ while holding r, $\dot{r}$ and $\dot{\theta}$ constant. As a consequence, (8.25b) reduces to

$$\frac{\mathrm{d}}{\mathrm{d}t}\left(\frac{\partial L}{\partial \dot{\theta}}\right) = 0, \tag{8.29}$$

and therefore

$$\frac{\partial L}{\partial \dot{\theta}} = \text{constant}. \tag{8.30}$$

But from (8.26) we find that $\partial L/\partial \dot{\theta} = mr^2\dot{\theta}$, so

$$r^2\dot{\theta} = \text{constant}. \tag{8.31}$$

This was a key result in Chapter 6, closely related to Kepler's second rule concerning planetary motion (see Section 6.3). There, we saw (8.31) as a consequence of the fact that the central force has, by definition, no transverse component. According to our 'new' view the quantity $r^2\dot{\theta}$ is conserved because the Lagrangian L does not depend explicitly on θ.

This lack of explicit dependence is called a *symmetry* of the system, and the connection between symmetries and conservation laws—of which we have just seen one example—is a recurring theme of modern theoretical physics.

Exercises

8.1 *The refraction of light*. Consider two different media separated by a plane boundary, and let the velocity of light be c_1 in the first medium and c_2 in the second. Show that if light travels from a given point A in the first medium to a given point B in the second medium *in the shortest possible time*, then

$$\frac{\sin \theta_1}{\sin \theta_2} = \frac{c_1}{c_2},$$

where θ_1 is the angle of incidence and θ_2 is the angle of refraction (Fig. 8.8). [In the refraction of light at the boundary of two given media the ratio $\sin \theta_1 / \sin \theta_2$ is indeed observed to be constant; this is *Snell's law* (1620).]

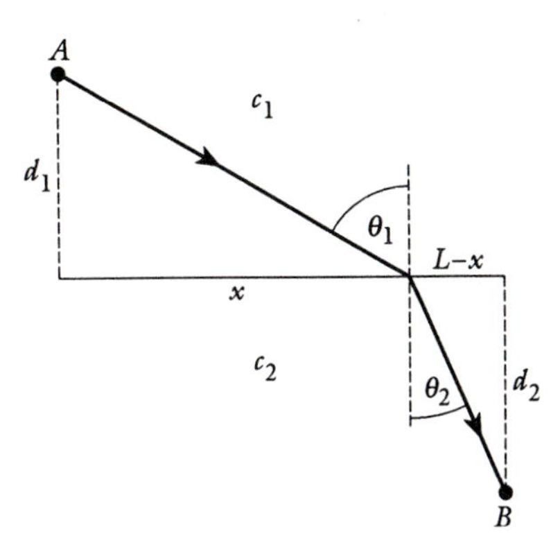

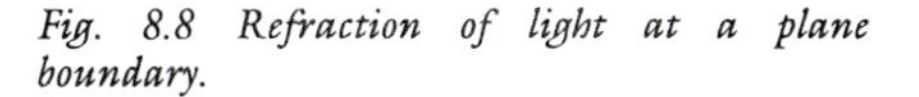

Fig. 8.8 Refraction of light at a plane boundary.

8.2 *A soap film problem*. Show that (8.15) reduces to

$$y\ddot{y} - \dot{y}^2 = 1,$$

and then solve this equation with the boundary conditions $y(-a) = y(a) = 1$ to obtain

$$y = c \cosh\left(\frac{x}{c}\right),$$

where the constant c is given by

$$c \cosh\left(\frac{a}{c}\right) = 1.$$

By considering the curves $z = \cosh \xi$ and $z = \xi / a$, or otherwise, show that the constant c is real only if $a < 0.6627$, i.e. if $a < 1/\sinh \xi_c$, where ξ_c is such that $\coth \xi_c = \xi_c$.

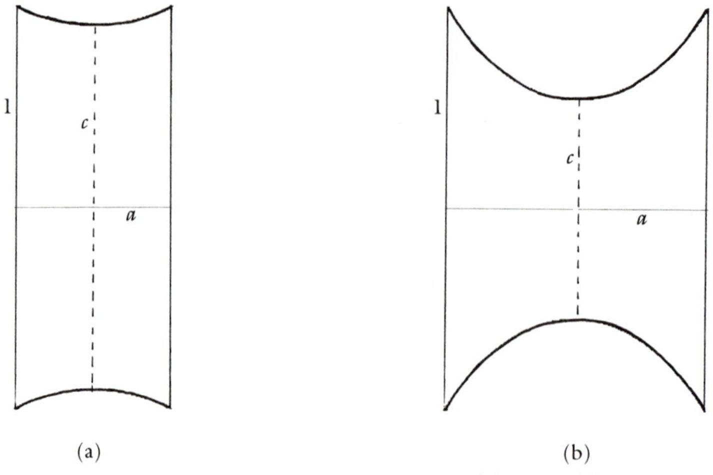

Fig. 8.9 The soap film spanning two circular rings. (a) $a = 0.4$, (b) $a = 0.6627$.

8.3 *Lagrange's approach to dynamics*. Use the method in Section 8.4 to obtain the equations of motion for:

(i) the simple pendulum of Fig. 5.1;
(ii) the double oscillator of Fig. 5.8;

noting that the elastic potential energy in a stretched linear spring is derived in Ex. 3.6.

9 Fluid flow

9.1 Introduction

At the very beginning of the twentieth century one of the major puzzles in dynamics concerned fluid motion.

To understand the problem we must note first that some fluids are less *viscous* than others. No reader will be surprised, for example, that the 'coefficient of viscosity' μ is much less for water than it is for syrup. Indeed, many fluids, such as water or air, hardly seem to be viscous at all, and it is then only natural to construct a simplified theory of fluid motion in which viscous effects are neglected altogether, with μ set equal to zero.

The theory of such 'non-viscous' fluid motion was initiated by Euler in the 1750s, and by the end of the nineteenth century it had accounted extremely well for such diverse flow phenomena as surface waves on water, sound waves in air, and even the motion of vortices such as smoke rings and tornados.

Yet, when the same non-viscous theory was applied to the streaming motion of air or water past a solid object, such as the circular cylinder in Fig. 9.1, it gave completely wrong answers, bearing no resemblance to the real

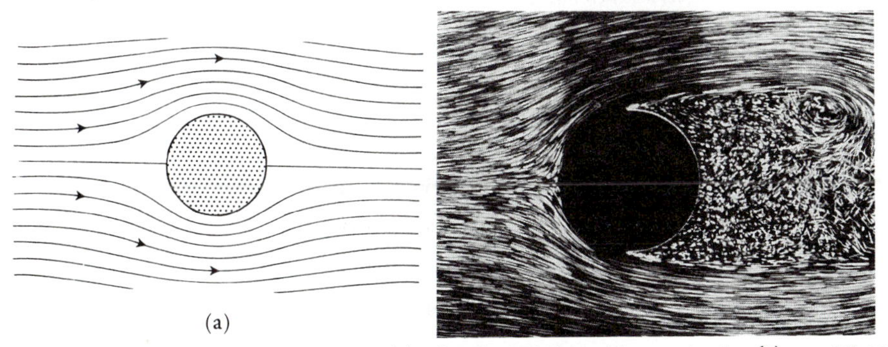

Fig. 9.1 Flow past a circular cylinder for (a) a hypothetical fluid with zero viscosity, (b) a real fluid with very small viscosity μ (from van Dyke 1982).

flow, especially downstream of the body. Moreover, this discrepancy seemed to persist *no matter how small the actual value of μ.*

In mathematical terms, then, the flow in the limit $\mu \rightarrow 0$ appeared to be quite different from the flow with μ *equal* to zero.

Why?

9.2 The geometry of fluid motion

Before we can consider the physical causes of fluid motion we must decide how to describe it in mathematical terms.

One natural possibility is to express the position (x, y, z) of any fluid 'particle' in terms of its original position (X, Y, Z) and time t. As a simple example, consider

$$x = X\,\mathrm{e}^{\alpha t}, \qquad y = Y\,\mathrm{e}^{-\alpha t}, \qquad z = Z, \tag{9.1}$$

where α is a positive constant. At $t = 0$ we have $(x, y, z) = (X, Y, Z)$, as required. As time goes on, the fluid is being continually 'stretched' in the x-direction and 'shrunk' in the y-direction, on account of the factors $\mathrm{e}^{\alpha t}$ and $\mathrm{e}^{-\alpha t}$. As $xy =$ constant for any given X, Y, the path of each fluid particle is a rectangular hyperbola (Fig. 9.2).

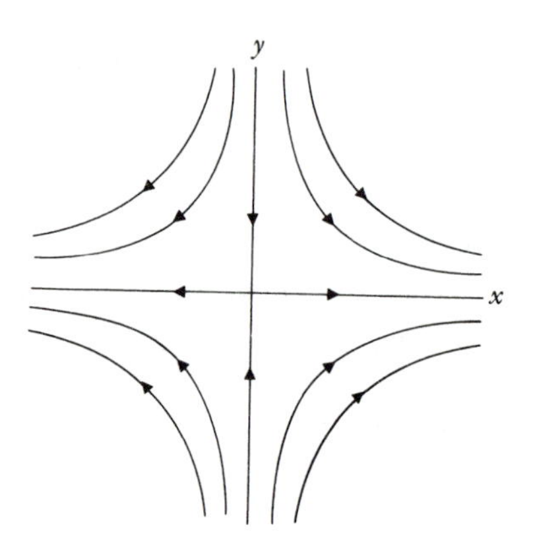

Fig. 9.2 The stretch-and-shrink motion of equations (9.1) and (9.3).

To obtain the *velocity* (u, v, w) of any particular fluid particle we simply differentiate partially with respect to t while holding X, Y, and Z constant. In the example above this gives

$$u = \alpha X\,\mathrm{e}^{\alpha t}, \qquad v = -\alpha Y\,\mathrm{e}^{-\alpha t}, \qquad w = 0. \tag{9.2}$$

This leads us to the most common and helpful way of describing a fluid flow, namely by means of an expression for the velocity (u, v, w) in terms of x, y, z, and t. On eliminating X, Y, and Z between (9.1) and (9.2) we obtain

$$u = \alpha x, \qquad v = -\alpha y, \qquad w = 0 \tag{9.3}$$

as a description of the flow in Fig. 9.2. It is this kind of representation which we shall use throughout the rest of the chapter.

Incompressible fluids

With many fluids, such as water or syrup, we want to express the idea that there can be effectively no change in the *volume* of any particular fluid blob as it deforms and moves about.

This might at first seem a very difficult task, but it can be shown that the partial differential equation

$$\frac{\partial u}{\partial x} + \frac{\partial v}{\partial y} + \frac{\partial w}{\partial z} = 0 \tag{9.4}$$

expresses just this idea, namely that the fluid is **incompressible**. We note at once that the flow given by (9.3) satisfies this equation, which is in keeping with our earlier remarks about the fluid being continually stretched in one direction and shrunk (equally) in another.

The quantity on the left-hand side of (9.4) is called the **divergence** of the velocity field. For a compressible fluid, such as air, a positive divergence indicates that a small fluid blob at the point (x, y, z) and time t in question is expanding in volume, while a negative divergence indicates that it is contracting.

Fictitious fluids

The above ideas are geometrical rather than physical, and one merit of this is that they may be used to advantage even in dynamical problems involving no 'real' fluid at all. It can then be helpful to *invent* one, instead, in *phase space* (Section 5.5).

Consider as an example the damped linear oscillator (5.14):

$$\ddot{x} + k\dot{x} + \omega^2 x = 0, \tag{9.5}$$

where $k > 0$. We may rewrite this as a pair of first-order equations, namely $\dot{x} = y$ and $\dot{y} = -ky - \omega^2 x$ (cf. (5.35)), and we may view this in turn as a *flow* in the phase plane with velocity field $u = \dot{x}$, $v = \dot{y}$, i.e.

$$\begin{aligned} u &= y, \\ v &= -ky - \omega^2 x. \end{aligned} \tag{9.6a,b}$$

The divergence of this flow is

$$\frac{\partial u}{\partial x} + \frac{\partial v}{\partial y} = -k, \tag{9.7}$$

which is negative, so there is continual contraction of the phase fluid. This is what we would expect; Ex. 5.1 shows that individual paths in the phase plane spiral into the origin in this case, corresponding to decaying oscillations in the real physical system (Figs. 5.18 and 9.3).

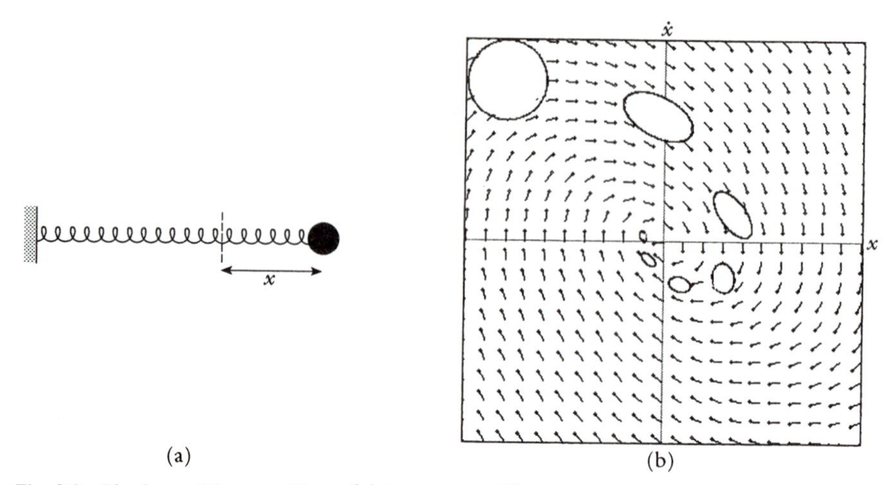

Fig. 9.3 The damped linear oscillator (a) in real space (b) in phase space, where a particular blob of 'phase fluid' can be seen contracting in volume as time goes on.

9.3 The equations of viscous flow

In order to proceed further with real fluid dynamics we must first make the idea of **viscosity** more precise.

To this end, consider the simple flow of Fig. 9.4, in which the velocity u of the fluid is in the x-direction and is a function of y only. Then the fluid just above some level $y =$ constant exerts a force on the fluid just below (and vice versa), and for many common fluids the tangential component of this force is found to be

$$\tau = \mu \frac{du}{dy} \tag{9.8}$$

per unit area of contact between the two parts of the fluid. This viscous force is therefore proportional to the velocity gradient, i.e. to the rate at which the fluid is being deformed, and the constant of proportionality μ is called the

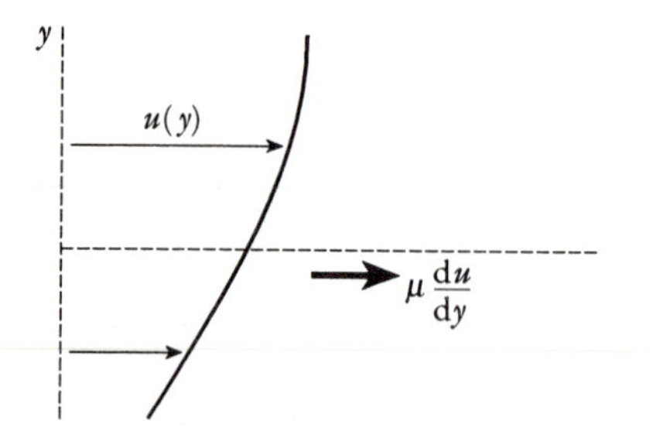

Fig. 9.4 The viscous stress in a simple shear flow.

Table 9.1 Typical values of the viscosity μ and density ρ for some common fluids.

	μ (kg m^{-1} s^{-1})	ρ (kg m^{-3})
Air	0.00002	1.3
Water	0.001	1000
Glycerine	1.8	1000
Golden Syrup	120	1000

coefficient of viscosity. Typical values of μ for some common fluids are given in Table 9.1.

The fundamental equations

The general equations of motion for a viscous fluid were obtained by Sir George Stokes in 1845. We shall concentrate on the special case when the fluid is incompressible and of constant density ρ, and we shall also suppose that the flow is steady and two dimensional, so that the velocity field is independent of time and of the form $[u(x, y), v(x, y), 0]$, as in the simple example (9.3).

In this case the equations of motion turn out to be

$$\rho\left(u\frac{\partial u}{\partial x} + v\frac{\partial u}{\partial y}\right) = -\frac{\partial p}{\partial x} + \mu\left(\frac{\partial^2 u}{\partial x^2} + \frac{\partial^2 u}{\partial y^2}\right),$$

$$\rho\left(u\frac{\partial v}{\partial x} + v\frac{\partial v}{\partial y}\right) = -\frac{\partial p}{\partial y} + \mu\left(\frac{\partial^2 v}{\partial x^2} + \frac{\partial^2 v}{\partial y^2}\right), \qquad (9.9\text{a, b, c})$$

$$\frac{\partial u}{\partial x} + \frac{\partial v}{\partial y} = 0,$$

where p denotes the pressure in the fluid.

Any reader who finds these equations daunting is in good company; Euler was appalled even by their *non*-viscous ($\mu = 0$) counterparts, which he derived in 1755. What matters for present purposes is just the overall structure of the equations (9.9).

Note, first, that we have the correct number of equations—three—for the three unknowns u, v, and p. We recognize the last of these, (9.9c), as expressing the incompressibility of the fluid (cf. (9.4)).

The first two equations are, in fact, just the x and y components of the statement 'mass × acceleration = force,' the right-hand sides representing a combination of pressure forces and viscous forces. The form of the acceleration terms on the left-hand side (in brackets) may seem a little unusual at first, but consider for a moment the much simpler problem of a single particle moving in, say, the x-direction. Its acceleration will be $\mathrm{d}u/\mathrm{d}t$, where $u = \mathrm{d}x/\mathrm{d}t$, so it can be written as $u\,\mathrm{d}u/\mathrm{d}x$. This is essentially where all the terms on the left-hand sides of (9.9a, b) are coming from.

Boundary conditions

A viscous fluid certainly cannot flow through a rigid impermeable boundary, but it is rather less obvious, perhaps, that it cannot slip along such a boundary, either. Experiments indicate that this second requirement, called the **no-slip condition**, must be satisfied by any viscous fluid, no matter how small its viscosity μ ($\neq 0$).

The two conditions, taken together, imply that at any point of a rigid boundary the velocity of the fluid must be equal to that of the boundary itself. In particular, we must have

$$u = v = 0 \quad \text{at a rigid boundary which is at rest} \tag{9.10}$$

(see Fig. 9.5).

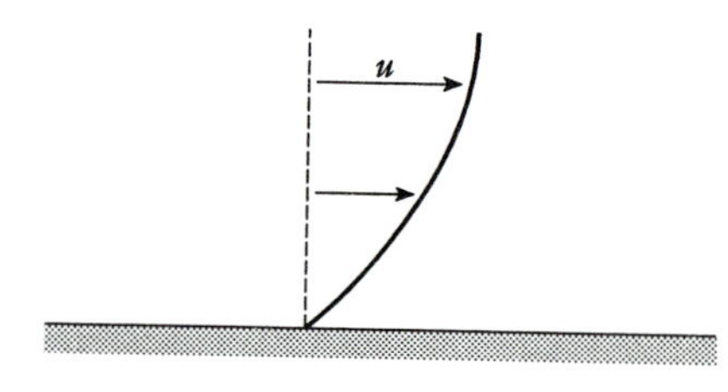

Fig. 9.5 Viscous flow near a rigid boundary, showing no slip.

The Reynolds number

In view of the difficulty of the full equations (9.9) it is natural to hope for circumstances in which some of the terms are much smaller than others, so that they may perhaps be neglected.

Now, if U denotes some typical flow speed and L denotes some typical length scale for the particular flow in question, it is not too difficult to argue that the acceleration terms in (9.9a, b) are *typically* of order $\rho U^2/L$, while the viscous terms are typically of order $\mu U/L^2$. These are only very rough estimates, but the ratio of the two is clearly

$$R = \frac{\rho U L}{\mu}, \tag{9.11}$$

which is called a **Reynolds number** for the flow.

So, *if* the above estimates are roughly correct, we may say that in (9.9a, b)

$$\frac{\text{acceleration terms}}{\text{viscous terms}} \approx R, \tag{9.12}$$

Table 9.2 Typical Reynolds numbers for various flows.

	R
Swimming spermatozoan ($U \sim 10^{-4}$, $L \sim 10^{-5}$)	10^{-3}
Golden Syrup draining from a spoon ($U \sim 10^{-2}$, $L \sim 3 \cdot 10^{-2}$)	2×10^{-3}
Finger moved through water ($U \sim 3.10^{-2}$, $L \sim 10^{-2}$)	3×10^{2}
Spin-down of a stirred cup of tea ($U \sim 10^{-1}$, $L \sim 5 \cdot 10^{-2}$)	5×10^{3}
Flow past wing of a Jumbo jet ($U \sim 200$, $L \sim 3$)	5×10^{7}

and we may then hope, at least, that one set of terms can be neglected in comparison with the other if either (i) R is very small or (ii) R is very large. We now consider these two limiting cases in turn.

9.4 Very viscous flow

Suppose that we have two circular cylinders, one inside the other, and that we fill the gap between them with Golden Syrup. Suppose, too, that we inject dye, marking one particular blob of syrup as in Fig. 9.6(a).

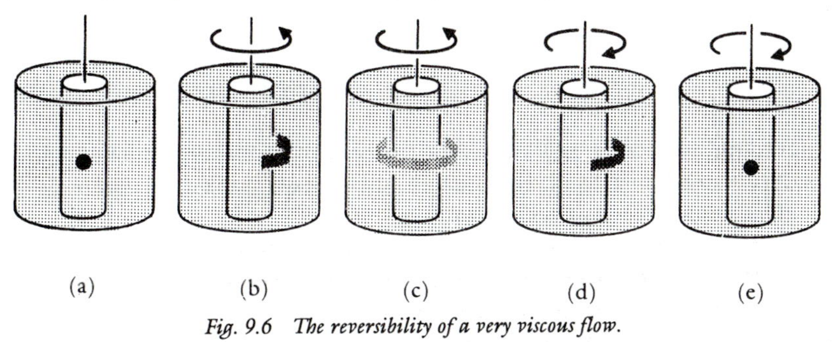

Fig. 9.6 The reversibility of a very viscous flow.

If we now slowly turn the inner cylinder, the fluid in contact with the cylinder will turn with it, because of the no-slip condition. Viscous forces within the fluid will then communicate some of this motion to the fluid lying further from the symmetry axis, and as a result the dyed blob of fluid will begin to get sheared, as in Fig. 9.6(b). After, say, four complete turns of the inner cylinder the blob will have become so drawn-out that it resembles a faintly dyed ring encircling the inner cylinder (Fig. 9.6(c)).

Remarkably, however, if we now slowly turn the inner cylinder back through four turns, to its original position, *the dyed portion of syrup will return almost exactly to its original state as a concentrated blob* (Fig. 9.6(d), (e)).

If this seems quite out of keeping with most everyday experience of fluid motion it is because most 'everyday' flows have a high Reynolds number. The present flow, however, has a very *low* Reynolds number, partly on account of the high viscosity of Golden Syrup (see Tables 9.1 and 9.2). And low Reynolds number flows are, indeed, almost *reversible*.

To see why this should be so, note from (9.12) that the acceleration terms on the left-hand sides of (9.9a) and (9.9b) appear to be negligible compared with the viscous terms in this limit, so that a low Reynolds number flow should be governed to good approximation by the equations

$$0 = -\frac{\partial p}{\partial x} + \mu\left(\frac{\partial^2 u}{\partial x^2} + \frac{\partial^2 u}{\partial y^2}\right),$$

$$0 = -\frac{\partial p}{\partial y} + \mu\left(\frac{\partial^2 v}{\partial x^2} + \frac{\partial^2 v}{\partial y^2}\right), \qquad (9.13\text{a, b, c})$$

$$\frac{\partial u}{\partial x} + \frac{\partial v}{\partial y} = 0.$$

Next, let u_1, v_1, p_1 be the solution to these equations satisfying the boundary conditions $u_1 = u_B$, $v_1 = v_B$ on some rigid boundary, where u_B and v_B are given. Now suppose that we 'reverse' those boundary conditions, so that we want $u = -u_B$, $v = -v_B$ on the rigid boundary instead. It is easy to confirm that the 'reversed' flow

$$u = -u_1, \qquad v = -v_1, \qquad p = \text{constant} - p_1 \qquad (9.14)$$

is the solution to our new 'reversed' problem; it clearly satisfies the new boundary conditions, and it satisfies the equations (9.13) because each individual term has *simply changed sign*.

This, then, is the explanation for the reversibility in Fig. 9.6. It is, emphatically, a low Reynolds number phenomenon, and viscous flows in general are most certainly not reversible; if we try replacing (u, v) by $(-u, -v)$ in (9.9) it typically will not work, because the viscous terms change sign but the acceleration terms—being quadratic in u, v—do not, so the 'reversed' flow is not a solution.

One exotic way in which this reversibility makes itself felt is in the swimming of micro-organisms such as spermatozoa. The Reynolds number R is, again, very small, essentially on account of the tiny length scale L (see Table 9.2). If these micro-organisms were to try to swim like fish, they would not get anywhere, because whatever they managed to achieve with one flap of the tail would be undone by the reverse flap which follows. Instead, they send helical waves down their tails (Fig. 9.7), which gets around the problem.

Fig. 9.7 A swimming spermatozoan.

9.5 The case of small viscosity

We now turn to the opposite extreme, in which μ is very small. We will assume, in other words, that the Reynolds number $R = \rho UL/\mu$ is very large, and this is certainly true for many flows of practical importance. Airflow past the wing of a passenger jet is typically characterized by Reynolds numbers of 10^7 or more; even the spin-down of a stirred cup of tea has $R \sim 10^4$ (Table 9.2).

When R is very large the viscous terms in (9.9a, b) *appear* to be absolutely negligible, on account of (9.12). Yet, as we pointed out in Section 9.1, there are circumstances in which, for some subtle reason, viscous effects *cannot* be neglected, no matter how large the value of R, i.e. no matter how small the value of μ.

The key to this puzzle lies in the fact that the viscous terms in (9.9a, b) are the ones *with the highest derivatives*, namely derivatives of second order. To see why this should be so important, consider the following much simpler illustrative problem.

A simple mathematical problem with a 'boundary layer'

Let $u(x)$ satisfy the ordinary differential equation

$$\varepsilon \frac{\mathrm{d}^2 u}{\mathrm{d}x^2} + \frac{\mathrm{d}u}{\mathrm{d}x} = 1, \tag{9.15}$$

subject to the boundary conditions

$$u(0) = 0, \qquad u(1) = 2, \tag{9.16}$$

where ε denotes a *very small* positive constant.

All that this little problem has in common with (9.9a, b, c) and their attendant boundary conditions is the way in which the highest derivative is multiplied by a very small coefficient; we are claiming no more than that the first term in (9.15) 'looks a bit like' the viscous terms in (9.9a, b), with ε taking the place of the small coefficient μ.

Now, if we attempt to neglect the first term in (9.15) on account of its very small coefficient, we quickly get into serious trouble, because we then have only a *first*-order differential equation

$$\frac{\mathrm{d}u}{\mathrm{d}x} = 1, \tag{9.17}$$

with *two* boundary conditions (9.16). The solution of (9.17) is $u = x + c$, and no choice of the constant c will allow us to satisfy *both* of those conditions; if we choose $c = 1$, for instance, so that $u(1) = 2$, then $u(0) \neq 0$.

Interestingly, we clearly run into this difficulty *no matter how small the value of* ε; even if ε were $10^{-1000000}$ we could still not neglect the first term in (9.15).

The great advantage of our simple problem over (9.9) is that we can solve it exactly:

$$u = x + \frac{1 - e^{-x/\varepsilon}}{1 - e^{-1/\varepsilon}} \tag{9.18}$$

(Ex. 9.3). We can then use this exact solution to track down the source of the difficulty.

Now, ε is greater than zero but very small, so $1/\varepsilon$ is very large and $e^{-1/\varepsilon}$ is extraordinarily small. The same may be said of $e^{-x/\varepsilon}$ over virtually all of the interval $0 \leq x \leq 1$, so

$$u \doteqdot x + 1 \tag{9.19}$$

there, as indicated in Fig. 9.8.

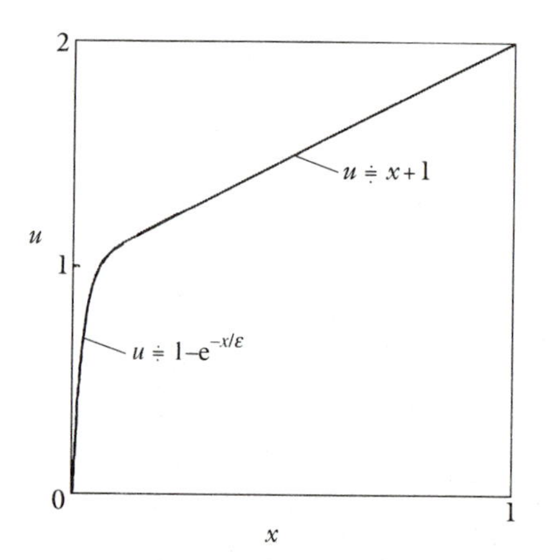

Fig. 9.8 The solution (9.18) when $\varepsilon = 0.02$.

But this argument clearly breaks down close to the $x = 0$ boundary, *when x becomes roughly as small as* ε, because neglecting $e^{-x/\varepsilon}$ in comparison with 1 then ceases to be valid. Indeed, as x decreases from, say, 5ε to 0, $e^{-x/\varepsilon}$ rises rapidly from about 0.01 to 1, so it eventually cancels the 1 in the numerator of (9.18).

Thus u is given by (9.19) virtually everywhere, but then changes very rapidly, from nearly 1 to precisely zero, in a very thin region adjacent to $x = 0$ called a **boundary layer** (see Fig. 9.8). In this boundary layer the derivative du/dx is large, of order $1/\varepsilon$, and d^2u/dx^2 is much larger still, of order

$1/\varepsilon^2$, as may be seen from the exact expressions

$$\frac{\mathrm{d}u}{\mathrm{d}x} = 1 + \frac{\frac{1}{\varepsilon}\,\mathrm{e}^{-x/\varepsilon}}{1 - \mathrm{e}^{-1/\varepsilon}}, \qquad \frac{\mathrm{d}^2u}{\mathrm{d}x^2} = -\frac{\frac{1}{\varepsilon^2}\,\mathrm{e}^{-x/\varepsilon}}{1 - \mathrm{e}^{-1/\varepsilon}} \tag{9.20a, b}$$

obtained from (9.18). This explains, finally, why we cannot neglect the first term in (9.15) entirely; it is *always* important in the boundary layer, and no matter how small we take ε the value of $\mathrm{d}^2u/\mathrm{d}x^2$ in the boundary layer will simply become larger in such a way that the term $\varepsilon\,\mathrm{d}^2u/\mathrm{d}x^2$ remains significant there.

Viscous boundary layers

Let us now return to real fluid flow at high Reynolds number. Figure 9.9 shows a fluid of small viscosity—water—flowing in a channel of decreasing cross-section. To visualize the flow, hydrogen bubbles have been released, in pulses, from a thin wire across the channel; this is roughly equivalent to marking small squares of fluid with dye.

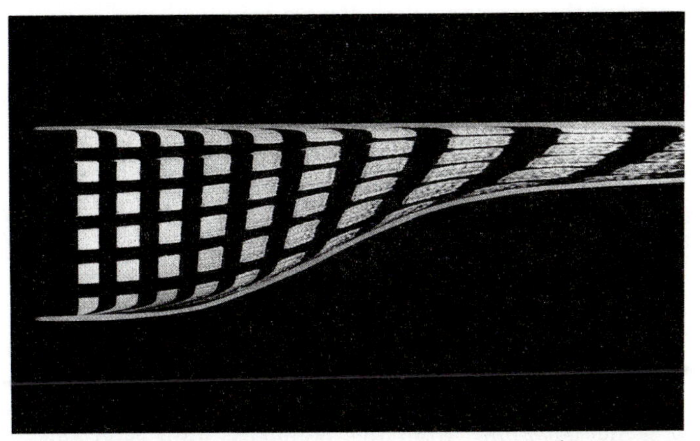

Fig. 9.9 Flow at high Reynolds number in a converging channel. (Encyclopaedia Britannica Educational Corporation).

And, if we look carefully, we can indeed see the squares behaving one way in the main body of the fluid and in a quite different way *in thin boundary layers close to each wall*, where viscous forces are important because of the exceptionally large velocity gradients (Fig. 9.10).

In this particular example the viscous boundary layers play a rather passive role, simply providing a rapid adjustment of the fluid velocity from its mainstream value to zero on the walls themselves, in accord with the no-slip

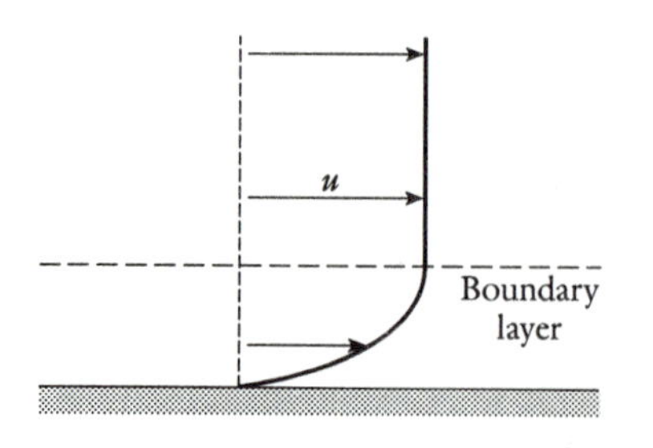

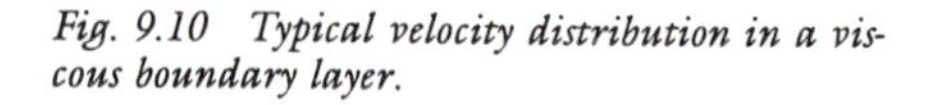
Fig. 9.10 Typical velocity distribution in a viscous boundary layer.

condition. More generally, however, viscous boundary layers can exert a controlling influence on the flow as a whole.

One example of this is the gradual spin-down of a stirred cup of tea, which is almost completely controlled by the thin viscous boundary layer on the *bottom* of the cup. To see this, take a flat-bottomed mug of cold water, throw in a few tea leaves as tracers, and stir briefly. The leaves can then be seen spiralling inwards in the bottom boundary layer, and this inward motion is crucial in driving a secondary circulation which is superimposed on the main rotary flow (Fig. 9.11). This in turn gradually changes a tall, thin blob in the main body of the fluid into a short, fat one. As a result, the angular velocity of the blob decreases, essentially by the same mechanism that ice-skaters use when putting out their arms to slow down a spin.

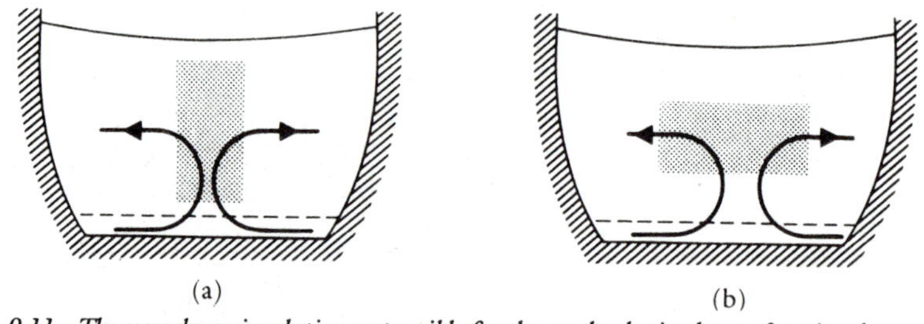

Fig. 9.11 The secondary circulation responsible for the gradual spin-down of a stirred cup of tea.

But the predominant way in which a thin boundary layer can control a main flow is by *separating* from the boundary itself, and this is what is happening in the flow past a circular cylinder shown in Fig. 9.1(b). The Reynolds number in this case is 2000, and the boundary layer on the upstream side of the cylinder is very thin, about $\frac{1}{50}$th of the cylinder radius. This thin layer of strongly sheared fluid then separates from the cylinder and so plays a crucial part in determining the flow downstream.

This whole phenomenon was first recognized by Ludwig Prandtl in an extraordinary eight-page paper published in 1905, and Fig. 9.12 shows his original sketch of the flow in a boundary layer near the separation point. He showed, too, that the thickness of the boundary layer is proportional to $\mu^{1/2}$, so that we may, in principle, make the boundary layer as thin as we like by

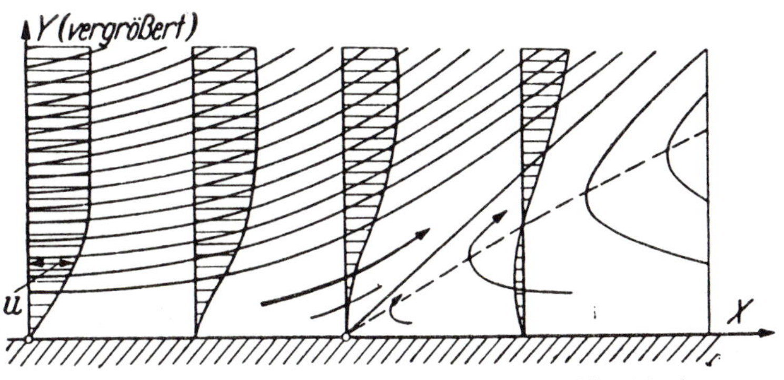

Fig. 9.12 The sketch of boundary-layer separation in Prandtl's original paper.

taking a fluid with sufficiently small μ. Significantly, however, he showed that we *cannot stop the separation* by doing this, even in the limit $\mu \to 0$.

In this way, then, Prandtl essentially resolved the deep puzzle mentioned in Section 9.1. Moreover, his ideas have subsequently found much wider application, way beyond the realms of fluid dynamics, for they show how it is possible, in principle, for an *arbitrarily small* physical cause to have a most significant effect.

In case this last remark should make viscous boundary layers and their separation seem at all rarefied, it is worth noting that without them, and the consequent shedding of a 'starting' vortex (Fig. 9.13), no aeroplane or bird would ever get off the ground.

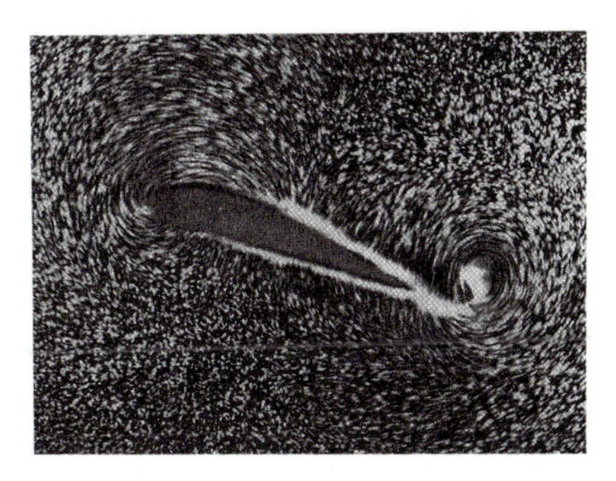

Fig. 9.13 Flow generated by suddenly moving a 'wing' to the left, through water which was initially at rest. Boundary layer separation at the sharp trailing edge, and the consequent shedding of a starting vortex, are crucial to the generation of 'lift'.

Exercises

9.1 Consider the flow

$$x = X \cos \Omega t - Y \sin \Omega t,$$

$$y = Y \cos \Omega t + X \sin \Omega t,$$

$$z = Z,$$

where Ω is a constant. Find the velocity components u, v, w in terms of x, y, z, and show that this particular motion would be a possible one for an incompressible fluid.

Show that each fluid particle moves in a circle, centred on $x = 0$, $y = 0$, and sketch the flow as a whole.

9.2 Show that the flow

$$u = \alpha x, \qquad v = -\alpha y, \qquad w = 0$$

in Fig. 9.2 is possible for an incompressible, viscous fluid, by showing that it satisfies the equations (9.9). What is the corresponding expression for the pressure p?

9.3 Show that the solution to the problem (9.15), (9.16) is (9.18), i.e.

$$u = x + \frac{1 - \mathrm{e}^{-x/\varepsilon}}{1 - \mathrm{e}^{-1/\varepsilon}}.$$

How does Fig. 9.8 change if ε is of very small magnitude but *negative*?

9.4 Solve

$$\varepsilon \frac{\mathrm{d}^2 u}{\mathrm{d}x^2} - u = -1,$$

where ε is a positive constant, subject to

$$u(0) = 0, \qquad u(1) = 0.$$

Sketch the solution when ε is very small.

10 *Instability and catastrophe*

10.1 Introduction

One of the most famous examples of **instability** occurs in an experiment first performed by Osborne Reynolds in 1883. He was studying the flow of water along a tube of circular cross-section, the water being drawn into the tube out

Fig. 10.1 Sketch of Reynolds' dye experiment, taken from his 1883 paper.

All this may be confirmed by direct numerical integration of (10.5), with or without a frictional damping term $-\tilde{k}\dot{\theta}$. A convenient way of doing this is to use the variation on the program `PENDANIM` discussed on pp. 225–6.

Stability of the 'new' equilibria

To show that the new equilibria are indeed stable we linearize (10.10) about $\theta = \theta_0$, writing

$$\theta = \theta_0 + \theta_1(t), \tag{10.13}$$

where θ_1 is assumed small.

As θ_0 is a constant, the left-hand side of (10.10) becomes $\ddot{\theta}_1$. If we write the right-hand side as

$$F(\theta) = \left[2(S-1)\cos\frac{\theta}{2} - S\right]\sin\frac{\theta}{2}, \tag{10.14}$$

then by Taylor's theorem (2.14)

$$F(\theta) = F(\theta_0) + \theta_1 F'(\theta_0) + O(\theta_1^2), \tag{10.15}$$

Now, $F(\theta_0) = 0$, because θ_0 satisfies (10.11), and on differentiating (10.14) as a product we obtain

$$F'(\theta_0) = -(S-1)\sin^2\frac{\theta_0}{2} + \left[2(S-1)\cos\frac{\theta_0}{2} - S\right]\tfrac{1}{2}\cos\frac{\theta_0}{2}.$$

The second term here is zero, again because θ_0 satisfies (10.11), so

$$\ddot{\theta}_1 = -(S-1)\sin^2\frac{\theta_0}{2}\cdot\theta_1 \tag{10.16}$$

is the linearized equation governing small displacements θ_1 about the new equilibrium positions $\theta = \theta_0$.

As these new positions given by (10.11) only exist if $S > 2$, the (constant) coefficient of θ_1 on the right-hand side of (10.16) is negative, and we have simple harmonic oscillations, showing that the 'new' positions are indeed stable.

10.4 Sudden changes of state

Generally speaking, if we gradually vary some parameter of a system we produce gradual variations in the state of the system itself. There can be circumstances, however, in which gradual variations of a parameter cause *sudden* or **catastrophic** changes of state.

Take, for example, a length of net-curtain wire, or bicycle brake cable.

Hold the wire firmly, perhaps with a pair of pliers, so that a short length L stands vertically upright, stabilized against gravity by its own elastic stiffness. Now gradually increase L.

Until some critical length $L = L_s$ the wire will spring back to the upward vertical even after a very large disturbance (Fig. 10.8(a)). As we continue to increase L we can keep the wire stable (Fig. 10.8(b)) until some second critical length $L = L_c$, provided we keep disturbances to the wire very small. As soon as L exceeds L_c, however, the smallest disturbance causes the wire to collapse to an entirely different 'flopped-down' state (Fig. 10.8(c)).

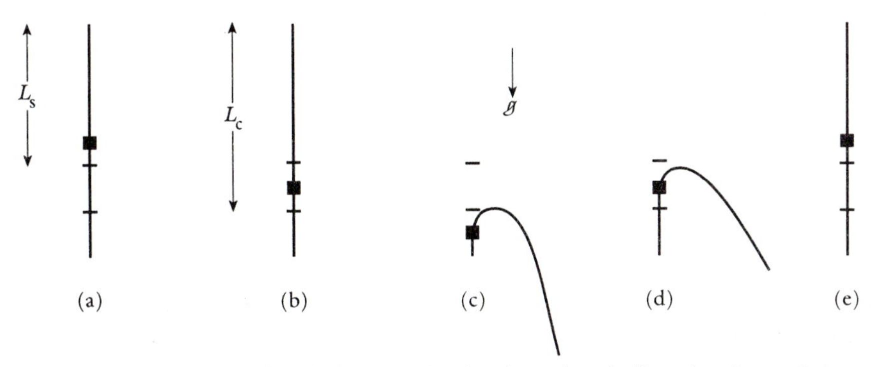

Fig. 10.8 Hysteresis with a length of net-curtain wire, clamped vertically at the point marked ■.

A small change in L can therefore cause a large *jump* in the state of this system. And there is more: if we gradually *decrease* L below L_c again, the system does *not* immediately jump back to its original state. Instead, the flopped-down state persists (Fig. 10.8(d)), provided we don't disturb it too much. Only as we decrease L back below the first critical length L_s does the wire spring back suddenly to the upward vertical of its own accord, so to speak.

The bifurcation diagram for a system of this kind is sketched in Fig. 10.9. Here A simply denotes some measure of the departure of the system from the equilibrium state $A = 0$, which is stable, according to linear theory, if some parameter L is less than a critical value L_c, and unstable if $L > L_c$. The bifurcation, or branching, of the solution at $L = L_c$ is said to be *subcritical*, in sharp contrast to the supercritical bifurcation in Fig. 10.6(a).

To check the diagram, suppose first that the system is in equilibrium with $A = 0$ and $L < L_c$. If we gradually increase L past L_c the $A = 0$ solution becomes unstable at the point Q in Fig. 10.9(b), and the system then *jumps* to either the state R or the state R', depending on which way it is 'nudged' by tiny disturbances. If it jumps to R, and we then gradually *reduce* L again, the 'new' solution persists until the point S, at which stage the system jumps back again to its $A = 0$ state at P.

This non-reversibility as a parameter is gradually increased and then

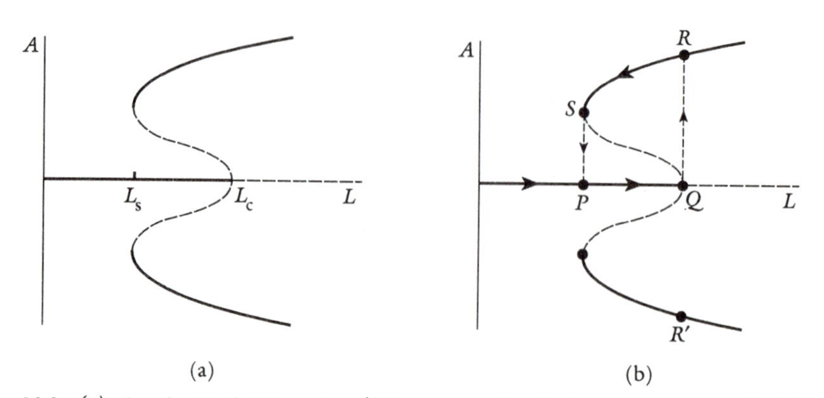

Fig. 10.9 *(a) A subcritical bifurcation (full curves stable, broken curves unstable). (b) An associated hysteresis loop.*

gradually decreased again is called **hysteresis**, and $PQRSP$ is called a hysteresis loop.

10.5 Imperfection and catastrophe

Let us return now to systems which have a *super*critical bifurcation as some parameter L is gradually increased past some critical value L_c (Fig. 10.10(a)). The system then gradually evolves along one of the two stable branches shown, which branch being determined by which way it is 'nudged' by small unavoidable disturbances when L is very slightly greater than L_c. If at some point we start to gradually decrease L again, the sequence of states reverses. There is, in particular, no sign whatever of the kind of 'jumpy' behaviour that we encountered in Section 10.4.

Now, this is all very well in theory, but in practice the branch that is followed as L passes L_c will often be determined not by small disturbances

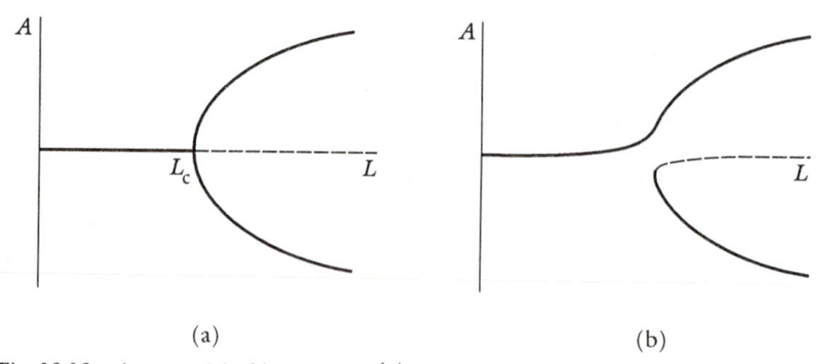

Fig. 10.10 *A supercritical bifurcation (a) and its modification by an imperfection (b).*

but by slight **imperfections** in the system itself which destroy the symmetry between the two 'new' solutions. The bifurcation diagram is then changed slightly, but crucially, to the form shown in Fig. 10.10(b) instead, and as we gradually increase L a definite branch is followed as L passes L_c. Moreover, if we then deliberately disturb the system so substantially that it settles down on the other branch, and we then gradually decrease L again, the system will *jump* back to the original state when L is small enough. The jump will be small if the asymmetry, or imperfection, of the system is small, but it will be there.

The classic Euler buckling experiment provides an example of this kind of behaviour, when we no longer have the perfect symmetry assumed in Fig. 10.4. Take a metal strip, or ruler, or even perhaps a suitable garden twig, provided it has a pronounced disposition to buckle one way rather than the other when compressed. Subject it to a substantial compressive force P, then press laterally on the middle of the strip so that it 'snaps' into its other, non-preferred state. On slowly decreasing P again there should, eventually, be a sudden and clearly observable jump back to the original state.

We now consider in detail a much simpler practical example of this kind of behaviour.

An example

Consider the system in Fig. 10.11. There, a light rod of length l is pivoted at one end O and has a point mass m at its other end. Its angle from the *upward* vertical is θ, and it is attached by two equal springs to a board BD which is at a given angle ε to the horizontal. The springs therefore tend to restore the rod to the position $\theta = \varepsilon$ somewhat *off* the upward vertical, so ε is the 'imperfection' which breaks the symmetry of this problem.

We assume that the force on the mass due to the springs is proportional to

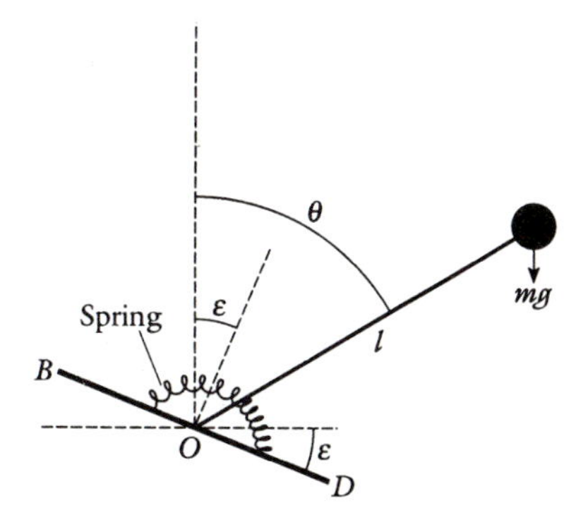

Fig. 10.11 A top-heavy rod with a spring attachment.

$\theta-\varepsilon$, and denote the constant of proportionality by $\bar{\alpha}$, which is a measure of the spring stiffness. The equation of motion is then

$$ml\frac{\mathrm{d}^2\theta}{\mathrm{d}t^2}=mg\sin\theta-\bar{\alpha}(\theta-\varepsilon). \tag{10.17}$$

If gravity were absent we would therefore have simple harmonic oscillations of the rod about $\theta=\varepsilon$ with frequency $(\bar{\alpha}/ml)^{1/2}$, proportional to the square root of the spring stiffness $\bar{\alpha}$.

Our interest, however, is in the case $g\neq 0$, and if we work with the dimensionless time variable

$$\tilde{t}=\left(\frac{g}{l}\right)^{1/2}t \tag{10.18}$$

then (10.17) becomes

$$\ddot{\theta}=\sin\theta-\frac{1}{M}(\theta-\varepsilon), \tag{10.19}$$

where a dot denotes differentiation with respect to $\tilde{t}$. Here

$$M=\frac{mg}{\bar{\alpha}} \tag{10.20}$$

is a dimensionless parameter representing the battle between the effect of gravity and the effect of the spring.

In order to simplify matters we shall consider only the case in which ε and θ are both fairly small, and we will approximate $\sin\theta$ in (10.19) by the first two terms of its Taylor series about $\theta=0$:

$$\sin\theta\doteqdot\theta-\tfrac{1}{6}\theta^3 \tag{10.21}$$

(see (2.15a)). This is a very good approximation for values of $|\theta|$ less than about 70° (see Fig. 2.6), and the equation of motion then becomes

$$\ddot{\theta}=\theta-\tfrac{1}{6}\theta^3-\frac{1}{M}(\theta-\varepsilon). \tag{10.22}$$

To obtain the equilibrium position(s) $\theta=\theta_0$ we set the right-hand side equal to zero, and this gives

$$(1-M)\theta_0+\tfrac{1}{6}M\theta_0^3=\varepsilon. \tag{10.23}$$

To see how θ_0 depends on ε it is rather easier to look at the problem the other way round, and plot ε as a function of θ_0. Clearly

$$\frac{\mathrm{d}\varepsilon}{\mathrm{d}\theta_0}=1-M+\tfrac{1}{2}M\theta_0^2, \tag{10.24}$$

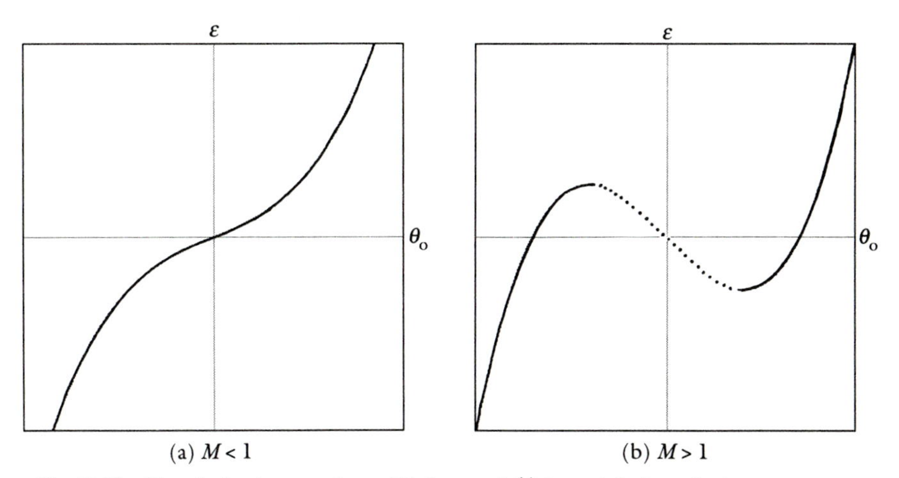

Fig. 10.12 The relation between the equilibrium angle(s) θ_0 and the imperfection parameter ε.

so if $M < 1$ the value of ε increases steadily with θ_0 (Fig. 10.12(a)), while if $M > 1$ there are stationary points at $\theta_0 = \pm 2^{1/2}(1 - M^{-1})^{1/2}$, and the curve takes the form shown in Fig. 10.12(b).

Finally, we turn the graphs of Fig. 10.12 through 90° and put them together in a three-dimensional way, so that the equilibrium angle(s) θ_0 can be seen as

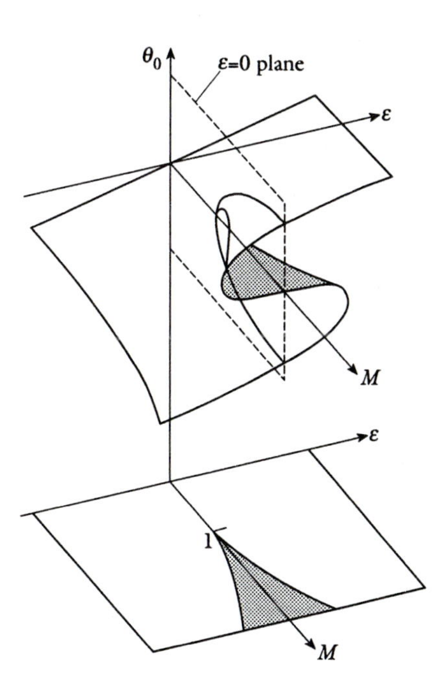

Fig. 10.13 Sketch of the equilibrium states as a function of both M and ε.

a function of both the mass parameter M and the imperfection parameter ε (Fig. 10.13). The key feature is the *fold* in the surface, and if we project this part down onto the M, ε plane we obtain the cusped region shown.

For any pair of values M, ε outside this region there is just one equilibrium position θ_0, but for any (M, ε) within the cusped region there are *three* equilibrium positions θ_0. The 'intermediate' one of these, corresponding to some point on the 'middle' (shaded) section of the fold in Fig. 10.13, can be shown to be unstable, while the other two are stable (Ex. 10.3). Bearing this in mind we can then, if we wish, fix M and follow how θ_0 changes as we gradually change the imperfection parameter ε. If $M > 1$, jumps in θ_0 can clearly occur, with associated hysteresis loops.

But the main value of Fig. 10.13 for present purposes is that we can now *fix* ε instead, and deduce the bifurcation curves giving θ_0 as a function of M. If $\varepsilon = 0$ we obtain in this way a perfectly symmetrical supercritical bifurcation at $M = 1$ of the kind shown in Fig. 10.10(a). With $\varepsilon > 0$, however, our simple physical example displays exactly the kind of behaviour shown in Fig. 10.10(b), with a 'disconnected' branch to the solution, and the rod having a definite tendency to lean one way rather than the other as the parameter M is gradually increased.

10.6 Instability of motion

We began, in Section 10.1, with a famous but very difficult example of the instability of a state of motion, rather than of a state of equilibrium. We now take a brief look at some simpler examples.

A misbehaving matchbox

Take an empty matchbox—preferably one for which all three linear dimensions are rather different from one another—and flick it up into the air, trying to make it spin about one of the three axes of symmetry through its centre (Fig. 10.14). You should find that you can make it spin successfully, i.e. stably, about the axes labelled 1 and 3, but that if you start it spinning about the axis labelled 2 the motion is unstable and the matchbox soon develops a violent wobbling motion of large amplitude.

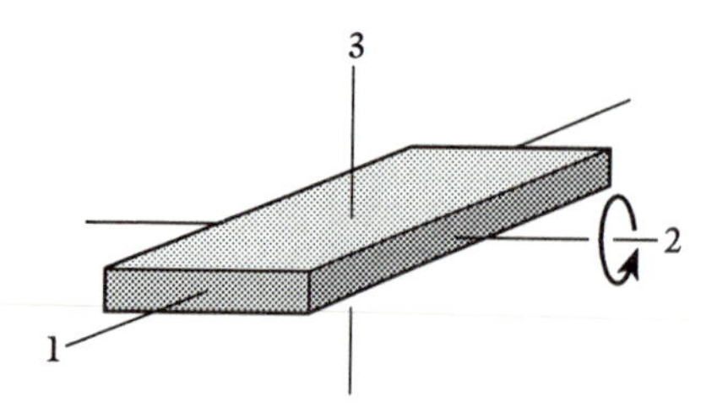

Fig. 10.14 A matchbox spinning (unstably) about one of its axes of symmetry.

To try to understand this, we turn to three beautifully symmetric equations of motion for a rigid object of any kind, first obtained by Euler in 1760:

$$A\dot{\omega}_1 = (B - C)\,\omega_2\,\omega_3,$$
$$B\dot{\omega}_2 = (C - A)\,\omega_3\,\omega_1, \qquad (10.25\text{a, b, c})$$
$$C\dot{\omega}_3 = (A - B)\,\omega_1\,\omega_2.$$

Here ω_1, ω_2, ω_3 denote the three components of the *angular velocity* of the object at any moment, measured with respect to a set of axes *fixed relative to the object*. (The symmetry axes of the matchbox in Fig. 10.14 certainly qualify here.) Thus if the object were simply rotating about the 1-axis we would have $\omega_1 = \omega$, say, and $\omega_2 = \omega_3 = 0$. In general, however, the angular velocity will have three components, and all three will be functions of time as the object wobbles about. The positive constants A, B, C are called the *moments of inertia* about the axes 1, 2, 3, respectively. The further from the axis in question that the mass is concentrated, the greater the moment of inertia, so for the case sketched in Fig. 10.14 we would have $A < B < C$.

Consider, then, the case when the matchbox is spinning about the 2-axis, with $\omega_2 = \omega$, where ω is a constant, and $\omega_1 = \omega_3 = 0$. This motion is certainly permissible according to the laws of dynamics, for it satisfies (10.25). But is it stable?

To answer this we perturb the motion a little, so that the angular velocity has components

$$\omega_1 = \xi_1, \qquad \omega_2 = \omega + \xi_2, \qquad \omega_3 = \xi_3 \qquad (10.26)$$

instead, where ξ_1, ξ_2, ξ_3 are all functions of time but *small* compared with ω. We then find from (10.25a, c) that

$$A\dot{\xi}_1 = (B - C)(\omega + \xi_2)\xi_3$$
$$C\dot{\xi}_3 = (A - B)\xi_1(\omega + \xi_2),$$

and on neglecting the quadratically small terms $\xi_2\,\xi_3$ and $\xi_1\,\xi_2$ we obtain the linearized equations

$$A\dot{\xi}_1 = (B - C)\,\omega\xi_3,$$
$$C\dot{\xi}_3 = (A - B)\,\omega\xi_1. \qquad (10.27\text{a, b})$$

Finally, we eliminate, say, ξ_3 by differentiating (10.27a), giving

$$AC\ddot{\xi}_1 = (A - B)(B - C)\,\omega^2\xi_1. \qquad (10.28)$$

As $A < B < C$ the (constant) coefficient of ξ_1 is positive, the equation is of the form $\ddot{\xi}_1 = p^2\xi_1$, and so we have instability (cf. (10.9)), as is observed in practice. If, on the other hand, B had been either the greatest or the least of the three moments of inertia, the factor $(A - B)(B - C)$ in (10.28) would have been negative, and ξ_1 would have oscillated in a simple harmonic way, corresponding to stability. This simple but elegant analysis is therefore fully in accord with the results of our casual experiment.

Spinning tops

Another curious example of unstable motion occurs with the so-called *tippy-top*, which is a child's toy consisting of a sphere with one segment chopped off and a stem attached in its place. The top will sit in stable equilibrium with its stem upright (Fig. 10.15(a)), but if we spin it sufficiently fast about the vertical the motion turns out to be unstable, and the top gradually turns itself upside-down (Fig. 10.15(b)). This particular instability mechanism is a very subtle one, with friction between the top and the underlying surface playing a crucial role.

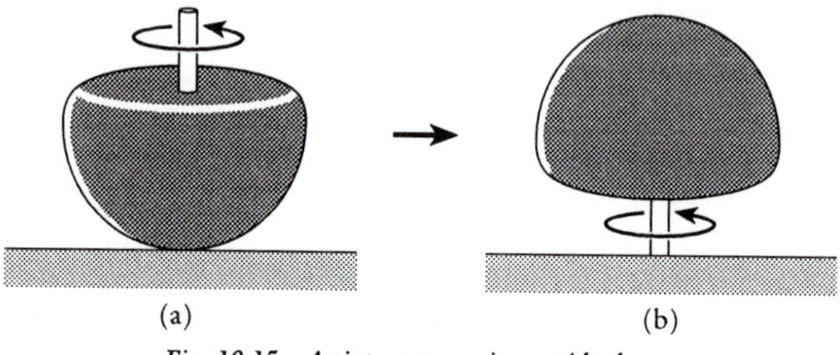

Fig. 10.15 A tippy-top turning upside-down.

The stability of state (b) is, however, no more mysterious than that of a conventional spinning top in its upright position (Fig. 10.16). Linear stability theory may again be brought to bear on the problem, and the upright state is found to be stable if the angular velocity of spin Ω exceeds some critical value Ω_c. The value Ω_c turns out to be greater for a tall, thin top than for a short, fat one, as we would expect.

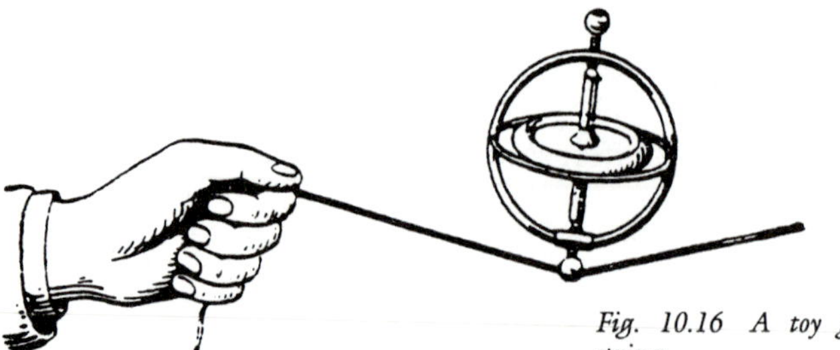

Fig. 10.16 A toy gyroscope balancing on a string.

The Reynolds experiment re-visited

One of the great difficulties with this particular experiment is that linear stability theory seems to predict that smooth, laminar flow down the tube in Fig. 10.1 is stable *at all speeds*.

There is no conflict here with the observations in Fig. 10.2(b), (c), for linear theory can guarantee stability only with respect to non-zero disturbances which are *sufficiently small* (so that the linearization procedure is valid). One of the troubles in the Reynolds system, apparently, is that at high enough flow speeds this stability would require a level of background disturbance so absurdly small as to be unachievable in any real experiment.

What really matters, in fact, is not the mean flow speed U itself but the dimensionless parameter

$$R = \frac{\rho U L}{\mu}, \tag{10.29}$$

now known as the Reynolds number (see Section 9.3). Here ρ is the density of the fluid, μ its coefficient of viscosity (see (9.8)) and L the diameter of the tube.

Recent experiments have shown that it is possible to keep the smooth, laminar flow stable up to values of R of 100,000 or so, but only by taking quite extraordinary pains to minimize disturbances, particularly those at the inlet to the tube. If we gradually increase R still further, the flow becomes turbulent, yet if we then start gradually *decreasing* R again the flow does not collapse back from its turbulent state to a laminar one until R is reduced

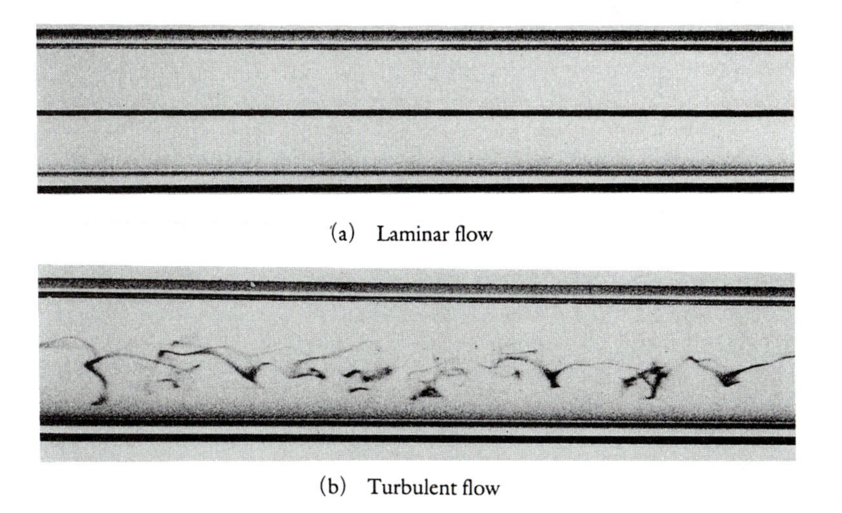

(a) Laminar flow

(b) Turbulent flow

Fig. 10.17 A repetition of Reynolds' experiment (from van Dyke 1982).

below a value of 2000 or so. The whole system therefore displays *hysteresis*, and in a much more extreme way than we encountered in Section 10.4.

Reynolds himself managed to keep the straight, laminar flow stable as far as $R \sim 13\,000$, and his original apparatus still stands in the hydraulics laboratory of Manchester University. A few years ago the experiment was repeated, on that same apparatus, in order to obtain the flow photographs in Fig. 10.17, but tiny vibrations from street traffic were enough to prevent the experimenters from doing as well as Reynolds himself in the horse-and-cart days of 100 years ago.

Exercises

10.1 *An elastic instability.* If we include friction in equation (10.5) we obtain

$$\ddot{\theta} = -\tilde{k}\dot{\theta} - \sin\theta + S\left(2\cos\frac{\theta}{2} - 1\right)\sin\frac{\theta}{2}$$

(cf. Ex. 5.5). To tackle this equation, adapt the animation program PENDANIM as suggested on p. 225. Explore the consequences of different initial values of θ and $\dot{\theta}$, and confirm that there are two stable, off-centre equilibria given by (10.11) when $S > 2$.

Let $S = 10$ and $\tilde{k} = 0.1$, and suppose that $\theta = 0$, $\dot{\theta} = 1.6$ at $\tilde{t} = 0$. At which of the two stable equilibria does the system eventually settle?

10.2 *Instability of a central-force orbit*. Suppose that a particle of mass m is moving under a central force c/r^n directed towards the origin O. Use (6.14) and (6.15) to show that

$$\ddot{r} - \frac{\hbar^2}{r^3} = -\frac{c}{mr^n},$$

where $\hbar = r^2\dot{\theta}$ is a constant. Deduce that uniform motion in a *circle* $r = a$ with angular velocity Ω is possible if $\Omega^2 = c/ma^{n+1}$.

Set $r = a + \eta(t)$, where η is small compared with a, and linearize the above differential equation about $r = a$. Deduce that uniform circular motion under a central force c/r^n is *unstable* if

$$n > 3.$$

10.3 *Multiple equilibria and catastrophe*. Consider the top-heavy, spring-supported rod in Fig. 10.11. If $|\theta|$ is reasonably small, equation (10.22) applies, and the equilibrium positions $\theta = \theta_0$ are given by $F(\theta_0) = \varepsilon$, where

$$F(\theta) = (1 - M)\theta + \tfrac{1}{6}M\theta^3$$

(see (10.23)).

Examine the stability of these equilibrium positions by linearizing (10.22) about $\theta = \theta_0$, and show, in particular, that equilibria along the full curves in Fig. 10.12 are stable, while those on the dotted part of the curve in Fig. 10.12(b) are unstable.

Show that when $M > 1$ sudden *jumps* in θ_0 occur as $|\varepsilon|$ is gradually increased beyond the critical value

$$|\varepsilon|_c = \left[\frac{8}{9}\frac{(M-1)^3}{M}\right]^{1/2},$$

which is therefore the equation of the cusped curve in the M, ε plane in Fig. 10.13. Confirm these jumps by using the variation on PENDANIM outlined on p. 226, which integrates the damped equivalent of (10.22).

10.4 *Hysteresis involving a state of motion.* If a damped simple pendulum is subject to a constant applied *torque* Γ, tending to turn the pendulum about its pivot, the dimensionless equation of motion may be shown to be

$$\ddot{\theta} + \tilde{k}\dot{\theta} + \sin\theta = \tilde{\Gamma},$$

(cf. Ex. 5.5), where

$$\tilde{\Gamma} = \frac{\Gamma}{mgl}.$$

Show that there are two equilibrium positions of the pendulum, one stable and one unstable, if $\tilde{\Gamma} < 1$, but that there are no equilibrium positions at all if $\tilde{\Gamma} > 1$.

Use the program derived from PENDANIM on p. 227 to show that as we gradually increase $\tilde{\Gamma}$ beyond the critical value 1 the pendulum suddenly switches from a state of equilibrium to a steady whirling motion about the pivot. Show, moreover, that if we then gradually *decrease* $\tilde{\Gamma}$ this whirling motion only collapses to give equilibrium again when $\tilde{\Gamma}$ falls below some substantially smaller critical value.

11 Nonlinear oscillations and chaos

11.1 Introduction

One of the 'simplest' dynamical systems exhibiting **chaos** is the equation

$$\ddot{x} + k\dot{x} + x^3 = A\cos\Omega t. \tag{11.1}$$

This is just a nonlinear counterpart to the classical forced oscillator of Section 5.2, with the restoring force due to the 'spring' being proportional to x^3 rather than to x itself. A typical numerical solution is shown in Fig. 11.1, with a fairly large forcing amplitude, $A = 7.5$.

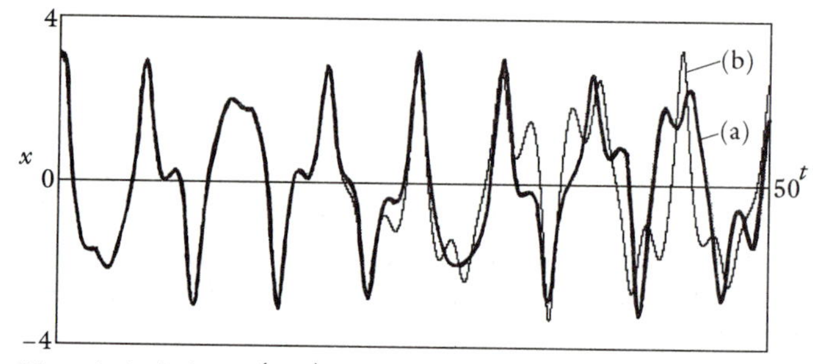

Fig. 11.1 Numerical solution to (11.1) with $k = 0.05$, $\Omega = 1$, $A = 7.5$ and initial conditions (a) $x = 3$, $\dot{x} = 3$, (b) $x = 3.003$, $\dot{x} = 3$.

Chaotic oscillations such as these have two essential features. First, the oscillations are persistently irregular, or haphazard, and never quite settle into a repeating pattern. Second, they display extreme **sensitivity to initial conditions**. The lighter curve in Fig. 11.1, marked (b), shows the effect of changing the initial conditions by just 1 part in 1000. At first, the two responses are practically indistinguishable, but after just 5 oscillation cycles of the driving term $A\cos\Omega t$ the two curves diverge rapidly into two quite different chaotic oscillations.

Within the last twenty years or so it has become clear that this kind of behaviour is entirely typical of many *nonlinear* dynamical systems, provided that they meet certain minimum requirements, which we discuss in Section 11.3.

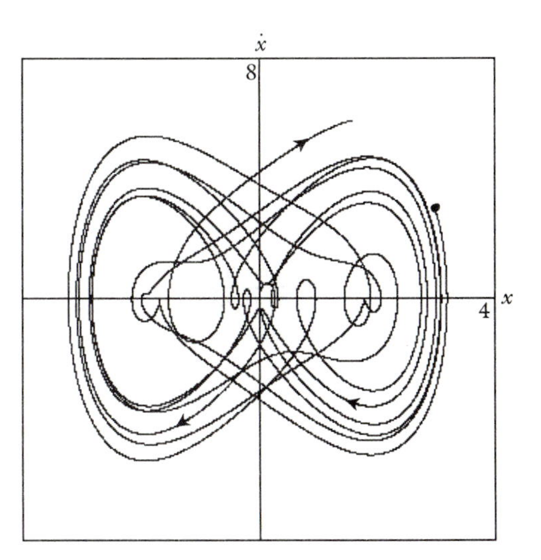

Fig. 11.2 The chaotic oscillation of Fig. 11.1(a) from a different perspective.

We should emphasize, however, that nonlinearity does not always act as an agent for chaos or disorder. Indeed, it can have precisely the opposite effect, and a good example of this follows in the next section.

11.2 Limit cycles; the van der Pol equation

The nonlinear system which we now consider has the intriguing property that if you do nothing to it initially, then nothing happens, but if you do *something* to it initially—no matter what—then the system ends up oscillating in one particular way, with an amplitude and frequency which is quite independent of the initial conditions.

The equation we have in mind arose, originally, in connection with a certain electrical circuit containing a triode valve (Fig. 11.3), and it is this

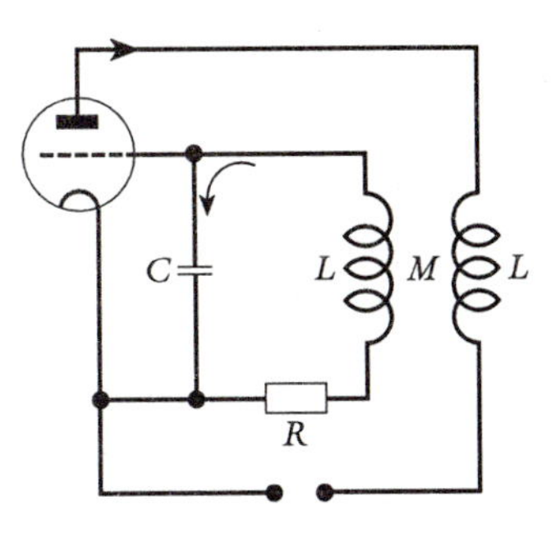

Fig. 11.3 The van der Pol circuit.

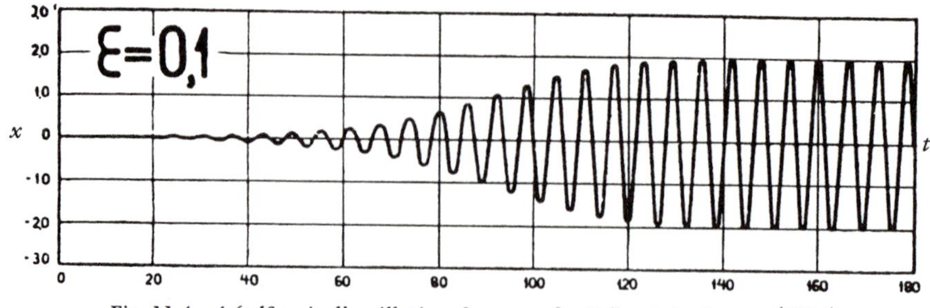

Fig. 11.4 *A 'self-excited' oscillation, from van der Pol's original paper (1926).*

valve which leads to the nonlinearity of the system. It turns out, in fact, that the non-dimensionalized electric current x is then governed by the so-called **van der Pol equation**

$$\ddot{x} + \varepsilon(x^2 - 1)\dot{x} + x = 0, \tag{11.2}$$

where ε is a positive constant.

Figure 11.4 shows, in the case $\varepsilon = 0.1$, the gradual onset of oscillations as a result of an initial value of x so small that it cannot be seen on the graph. And, as we have already intimated, an enormous initial value for x (or for $\dot{x}$) would still result, eventually, in the same outcome, namely the periodic oscillation to be seen for $t \gtrsim 140$ in Fig. 11.4. The amplitude of this oscillation is almost equal to 2, and the oscillations are almost simple harmonic, with period approximately 2π.

We can see, roughly, how such 'self-excited' oscillations come about by comparing (11.2) with the linearly damped oscillator equation (5.14). The coefficient of $\dot{x}$ in (11.2) is not constant, of course, but it is positive when $|x| > 1$, so the term as a whole might be expected to act as a damping mechanism when $|x|$ is reasonably large. When $|x|$ is small, on the other hand, the coefficient of $\dot{x}$ in (11.2) is negative, and we expect the term as a whole to have the opposite effect, causing *growth* of an oscillation, as would a negative value of k in (5.14) (see (5.15)).

If we now consider the phase plane, writing (11.2) in the form

$$\begin{aligned} \dot{x} &= y \\ \dot{y} &= -x - \varepsilon(x^2 - 1)y \end{aligned} \tag{11.3a,b}$$

(see Sections 3.6 and 5.5), we find that phase paths spiral *outward* from the neighbourhood of the origin, and one of them is shown in Fig. 11.5, which is for the case $\varepsilon = 1$. Two other paths, both starting from 'large' initial conditions, are also shown. All paths approach the closed curve (bold) which is called a **limit cycle**. At this value of ε it is distinctly non-circular, indicating that while the limit cycle is certainly a periodic oscillation it is no longer a simple harmonic one (cf. Fig. 5.15).

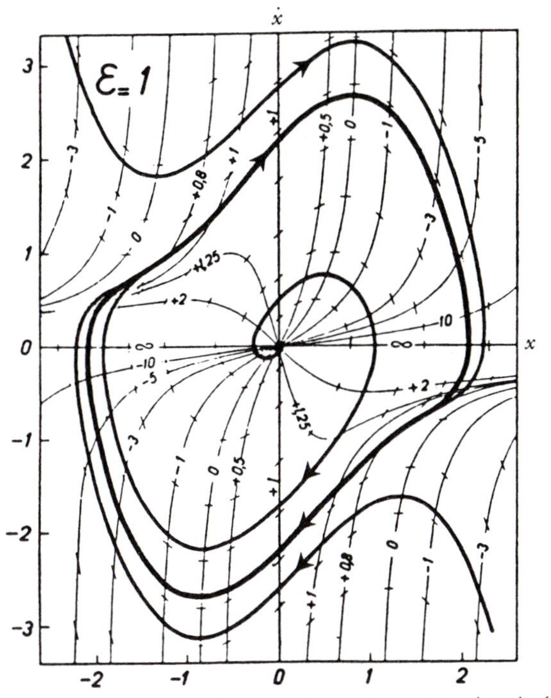

Fig. 11.5 Approach to the limit cycle in the phase plane, with $\varepsilon = 1$ in (11.2). (Some arrows have been added to van der Pol's original figure.)

Relaxation oscillations

For still larger values of ε the whole character of the limit cycle loses any resemblance to simple harmonic motion. In each cycle x decreases very slowly from the value 2 to the value 1 and then suddenly switches to the value -2 in a comparatively short time. During the second half of the cycle x increases slowly from -2 to -1, and then suddenly switches back to 2 so that the cycle can repeat. The larger ε, the more sudden these switches become, while the period of the oscillation as a whole gets longer and longer (Ex. 11.1).

The general behaviour shown in Fig. 11.6 is not uncommon in nonlinear systems which have a large (or small) parameter, and it is known as a *relaxation oscillation.*

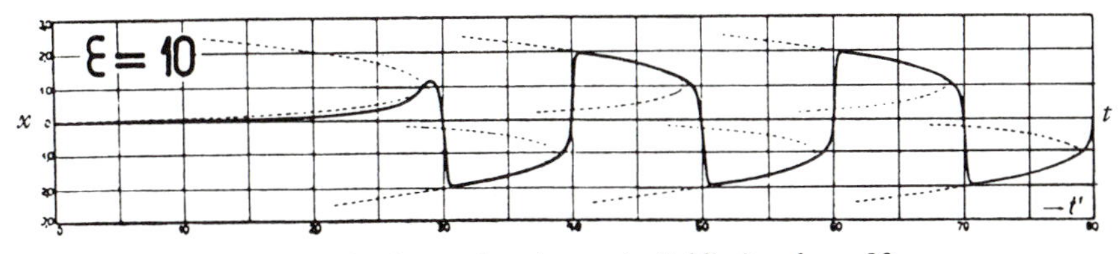

Fig. 11.6 Approach to the van der Pol limit cycle, $\varepsilon = 10$.

11.3 Conditions for chaos

We now return to the idea of *chaotic* oscillations in nonlinear systems. There is one key condition which must be met for such oscillations to be at all possible: the associated phase space *must be at least three-dimensional.*

To put the matter another way, chaos *cannot* occur in a two-dimensional autonomous system

$$\begin{aligned} \dot{x} &= f(x, y), \\ \dot{y} &= g(x, y). \end{aligned} \qquad (11.4\text{a, b})$$

The reason for this is an important result called the **Poincaré–Bendixon theorem**, which we now describe.

Fig. 11.7 Henri Poincaré (1854–1912).

Imagine first that we have determined the *equilibrium points*, if any, of the above system (11.4), i.e. the points at which both $f(x, y) = 0$ and $g(x, y) = 0$. Now suppose that a phase path starts at some point and cannot leave a certain bounded region of the x, y plane. Then the Poincaré–Bendixon theorem says that the phase path must eventually either (a) terminate at an equilibrium point, *or* (b) return to the original point, giving a closed path, *or*

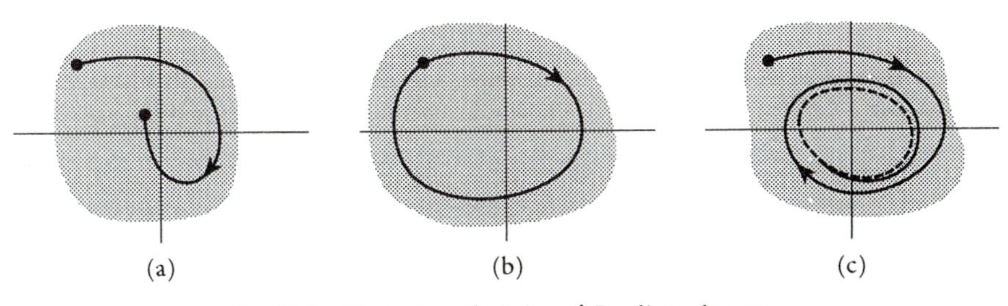

Fig. 11.8 Illustrating the Poincaré-Bendixon theorem.

(c) approach a limit cycle (Fig. 11.8). In particular, then, chaotic solutions are ruled out.

While the above theorem is certainly not obvious, one severe constraint which helps to bring the theorem about can be seen quite easily. For it follows immediately from (11.4) that

$$\frac{dy}{dx} = \frac{g(x,y)}{f(x,y)}, \tag{11.5}$$

which assigns a *unique* value of the phase path slope dy/dx to each point (x,y) of the phase plane, unless the point in question happens to be an equilibrium point, at which both $f(x, y)$ and $g(x, y)$ are zero. So a phase path in a two-dimensional autonomous system *can only cross itself, or another such path, at an equilibrium point*. This prohibits, at a stroke, a system with a *two*-dimensional phase space behaving in the kind of way shown in Fig. 11.2, with one of its phase paths crossing itself again and again in more and more different places as time goes on.

Note at once that the system in Fig. 11.2 is *not* of the form (11.4); the governing equation, (11.1), is second order but *not autonomous*. If we recast it in the form of an autonomous first-order system

$$\begin{aligned} \dot{x} &= y \\ \dot{y} &= -ky - x^3 + A\cos\Omega t \\ \dot{t} &= 1 \end{aligned} \tag{11.6}$$

(see Section 3.6), we see that the associated phase space is *three*-dimensional, so the Poincaré-Bendixon theorem does not apply. Note too that we are seeing in Fig. 11.2 only *the projection onto the x, y plane* of a chaotic phase path which rises steadily up out of the page—i.e. in the positive t-direction—and never in fact crosses itself at all.

It can happen, of course, that the phase space of a system is more 'obviously' three-dimensional from the very outset, and a famous example follows in the next section.

11.4 The Lorenz equations

These equations first arose in 1963 from a drastically over-simplified model of thermal convection in a layer of fluid. In their 'usual' form they are

$$\begin{aligned}\dot{x} &= 10(y - x),\\ \dot{y} &= rx - y - zx,\\ \dot{z} &= -\tfrac{8}{3}z + xy,\end{aligned} \tag{11.7a,b,c}$$

where r is a constant parameter.

In the original context, r acted as a measure of the imposed temperature difference between the bottom of the fluid layer and the top, which is what was driving the convective motion. In the same vein, x measured the flow speed, while y and z denoted certain broad features of the temperature distribution. Subsequently, however, these equations have attracted most attention purely on their own mathematical merits, and this is how we shall treat them here.

The natural first step is to find the equilibrium points of (11.7) and determine whether they are stable or unstable (see Sections 10.2, 10.3). The origin

$$x = 0, \qquad y = 0, \qquad z = 0 \tag{11.8}$$

is clearly an equilibrium point for all r, but it turns out to be stable according to linear theory only for $r < 1$. If we increase r beyond 1 we find two 'new' equilibrium points

$$x = y = \pm\sqrt{\tfrac{8}{3}(r-1)}\,, \qquad z = r - 1. \tag{11.9}$$

These exist for all $r > 1$ but turn out to be linearly stable only for $1 < r < 24.74$. No other equilibrium points exist.

In Fig. 11.9 we show some typical numerical solutions of (11.7) in the case $r = 28$, obtained using the program NXT. The chaotic nature of these solutions

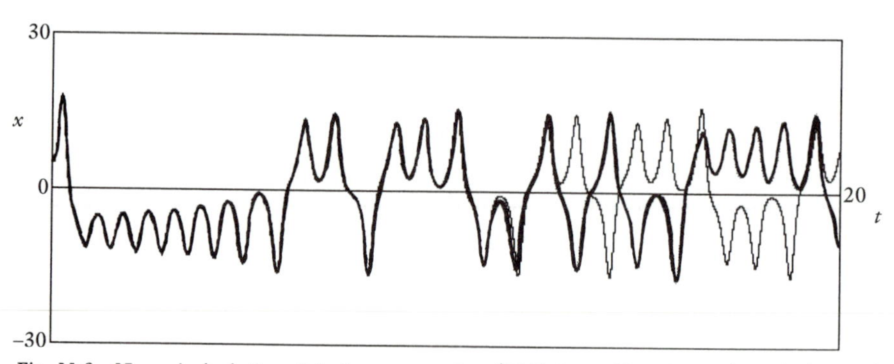

Fig. 11.9 Numerical solution of the Lorenz equations (11.7) for r=28 and with $(x, y, z) = (x_0, 5, 5)$ at t=0. The bold curve is for x_0=5.000, the lighter curve for x_0=5.005.

is evident, not just because of their irregularity, but because of their extreme sensitivity to initial conditions. With an initial difference of just 1 part in 1000 the oscillation sequences are seen diverging as t becomes greater than about 13. Moreover, even if we reduce the discrepancy in the initial conditions by a factor of 100, to just 1 part in 100 000, we only manage to keep the solutions together for a little longer, till t is about 16, after which they once again go their separate ways.

Lorenz saw this behaviour to be a general property of irregular oscillations in nonlinear systems; indeed, he realized that this extreme sensitivity to the initial conditions was essentially the *cause* of the irregularity. He realized, too, the practical implications, remarking in his 1963 paper that

> ...When our results...are applied to the atmosphere,...they indicate that prediction of the sufficiently distant future is impossible by any method, unless the present conditions are known exactly. In view of the inevitable inaccuracy and incompleteness of weather observations, precise very-long-range forecasting would seem to be non-existent.

Not surprisingly, chaotic oscillations are also sensitive to the errors introduced at each time step due to the approximate nature of any numerical method, and even greater care than usual is needed in that respect. The calculations in Fig. 11.9 were performed with double precision arithmetic and a Runge–Kutta routine with time step $h = 0.001$, followed by repeat calculations with $h = 0.002$ and $h = 0.0005$ as a check. A significantly larger step size, such as $h = 0.01$, did not permit the oscillations to be calculated correctly, even on the rather limited time scale $0 < t < 20$ shown in Fig. 11.9.

Some further understanding of these chaotic oscillations can be gleaned by following the path of a particular point (x, y, z) in phase space, as it evolves

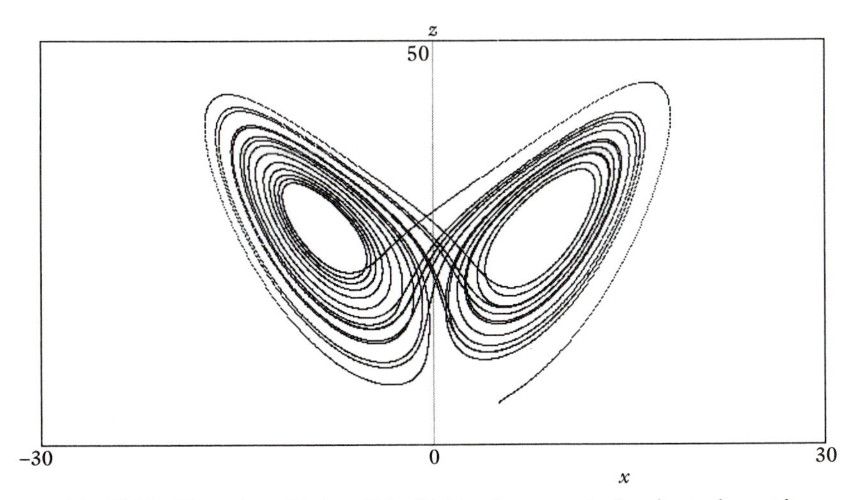

Fig. 11.10 The main oscillation of Fig. 11.9 in phase space, projected onto the x-z plane.

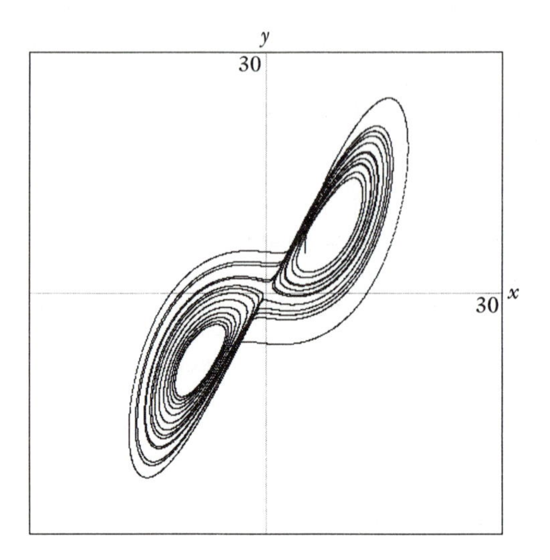

Fig. 11.11 Another view of the phase path in Fig. 11.10.

according to (11.7). Figure 11.10 shows the projection onto the x, z plane of the path corresponding to the initial conditions $(5, 5, 5)$ in Fig. 11.9. Despite appearances from this projection, the actual path in three dimensions does *not* intersect itself at all, and to help visualize it better it can be helpful to think of two gramophone records, held roughly at right angles and centred on the two equilibrium points (11.9). The phase point moves quickly from $(5, 5, 5)$ to a point on the left-hand 'record', spirals outward along a 'groove', then rather suddenly leaves that record and moves quickly to some point of the other, where the process repeats. Yet this simple analogy disguises, in fact, the full horrifying complexity of the situation, for the so-called **strange attractor** in Fig. 11.10 is not so much like two gramophone records as like two infinite stacks of infinitely-thin gramophone records, all grooved slightly differently.

One simple indication that something very strange is happening comes from an idea introduced under the heading 'fictitious fluids' in Section 9.2. If we view (11.7) as a fluid flow in phase space, writing $u = 10(y - x)$ etc., the divergence of the flow is

$$\frac{\partial u}{\partial x} + \frac{\partial v}{\partial y} + \frac{\partial w}{\partial z} = -10 - 1 - \tfrac{8}{3} = -13\tfrac{2}{3}, \tag{11.10}$$

which is negative.

Consider, then, a small blob of phase fluid initially centred on $(5, 5, 5)$. Because this represents a whole set of slightly different initial conditions, and because we know the outcome to be sensitive to these, we know that the blob will become greatly deformed and spread about all over the attractor in Fig. 11.10 in quite a short time. Yet because of the result (11.10) the blob must manage to do this while *decreasing in volume* all the time. Moreover, a divergence of −13.667 corresponds to a quite spectacular rate of shrinking; with each additional oscillation in Fig. 11.9 the volume of a blob of phase fluid decreases by another factor of 14 000 or so.

11.5 Chaotic mixing: stretch and fold

A third-order system of equations devised in 1976 by Otto Rössler gives us a clearer opportunity still to obtain some geometric insight into how chaotic motions can come about. In their usual form these equations are

$$\begin{aligned}\dot{x} &= -y - z,\\ \dot{y} &= x + 0.2y, \\ \dot{z} &= 0.2 + (x - c)z,\end{aligned} \qquad (11.11\text{a, b, c})$$

where c is a constant parameter. In contrast to the Lorenz equations they have just one nonlinear term, namely xz in (11.11c).

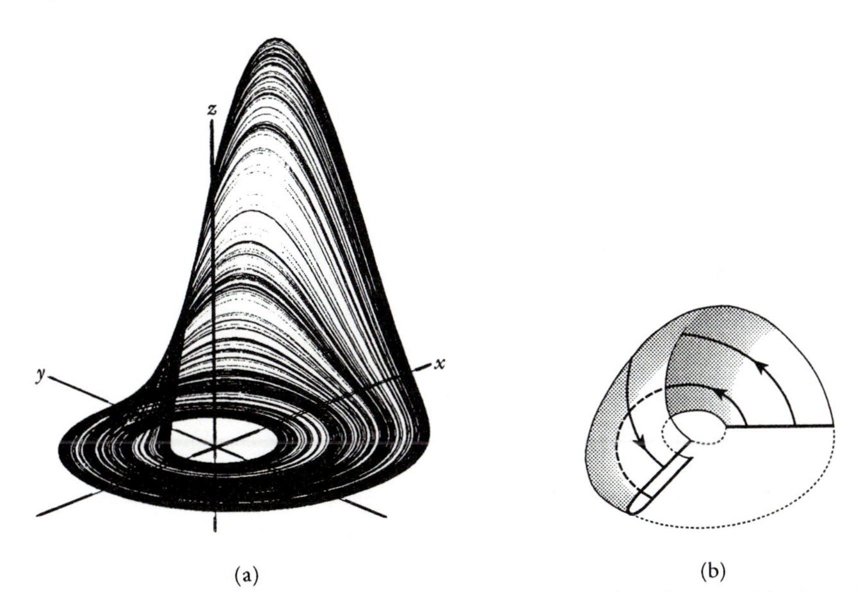

Fig. 11.12 (a) A typical chaotic trajectory for the Rössler equations (11.11); c=5.7. (b) Schematic picture of the repeated stretch-and-fold action of the phase flow. (After Peitgen et al. 1992.)

A typical chaotic phase path for $c = 5.7$ is shown in Fig. 11.12(a). For substantial periods of time z is very small, and x and y then evolve approximately according to the simplified equations

$$\begin{aligned}\dot{x} &= -y,\\ \dot{y} &= x + 0.2y,\end{aligned}$$

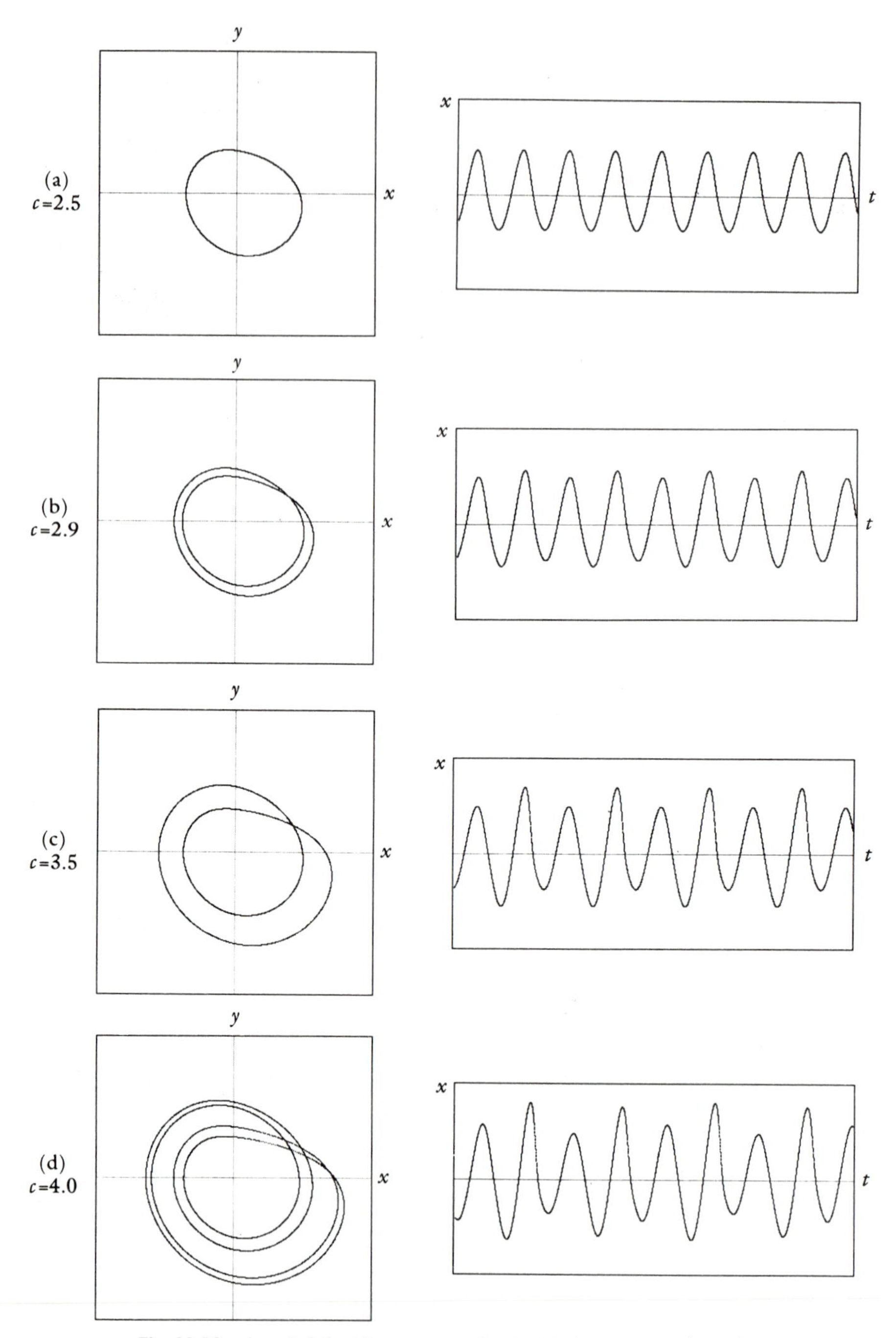

Fig. 11.13 A period-doubling sequence for the Rössler equations (11.11).

so that

$$\ddot{x} - 0.2\dot{x} + x = 0, \tag{11.12}$$

which corresponds to a growing oscillation (cf (5.14)), with an outward-spiralling path in the x, y plane. Eventually, however, x becomes greater than c, and (11.11c) then implies roughly exponential growth for z, so that the phase path turns sharply upward and out of the x, y plane. And as soon as z becomes large, the $-z$ term in (11.11a) makes $\dot{x}$ large and negative, which reduces x below c again, so that the whole process can continue.

In this way it is not too difficult to visualize how the flow in phase space manages to repeatedly stretch *and fold* a portion of surface (Fig. 11.12(b)) so that points corresponding to neighbouring initial conditions quickly find themselves being thoroughly dispersed, even within a distinctly finite region of phase space. This stretch-and-fold mechanism seems, more generally, to be the source of that *sensitivity to initial conditions* which we have already noted as one of the hallmarks of chaos.

11.6 One route to chaos: period-doubling

Systems which exhibit chaotic oscillations typically do so for some ranges of the relevant parameters but not for others, so one matter of obvious interest is how the chaos appears (or disappears) as one of the parameters is gradually varied. We may investigate this—up to a point—with the program NVARY.

The Rössler equations (11.11) provide a good example of a common route to chaos called **period-doubling**. Suppose, for instance, that we start with $c = 2.5$. For a wide range of initial conditions we end up eventually with the simple limit cycle oscillation shown in Fig. 11.13(a). If we then gradually increase c there are slow and unremarkable changes in the amplitude and period of oscillation until c increases beyond about 2.83. At that point the path in the phase plane just fails to close up on itself after one circuit, and does so instead after *two* circuits (Fig. 11.13(b)). By $c = 3.5$ this has become rather more pronounced (Fig. 11.13(c)), and at $c = 3.84$ the same thing happens again, leading to an oscillation pattern which only repeats itself after *four* cycles (Fig. 11.13(d)). As c is increased still further these period-doublings occur ever more frequently until at $c \sim 4.2$ the oscillation pattern is chaotic, never exactly repeating itself.

11.7 Multiple solutions and 'jumps'

We end this chapter by noting that a nonlinear system may well be capable of oscillating in several different ways *even for the same set of parameters*. The initial conditions will then determine which type of oscillation is actually observed.

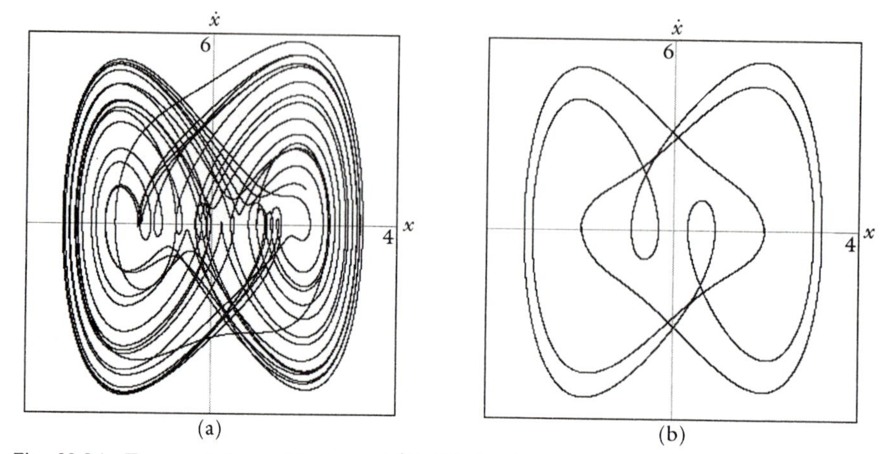

Fig. 11.14 Two coexisting oscillations of (11.13) for the same parameter values $k=0.2$, $\Omega=1$, $A=8.0$. (a) Chaotic, obtained with $x=1$, $\dot{x}=0$ at $t=0$. (b) Regular, period 6π, obtained (after transients have died down) with $x=0$, $\dot{x}=0$ at $t=0$.

The forced cubic oscillator provides a case in point:

$$\ddot{x} + k\dot{x} + x^3 = A \cos \Omega t \tag{11.13}$$

(see Section 11.1), and Fig. 11.14 shows a chaotic solution and a regular oscillation with frequency $\frac{1}{3}\Omega$ coexisting for the same set of parameters. In Ex. 11.5 we find an example of *five* different limit cycles coexisting, again, for one particular set of parameters.

An early example of such non-uniqueness arose in 1918 in connection with the **forced Duffing equation**

$$\ddot{x} + k\dot{x} + \alpha x + \beta x^3 = A \cos \Omega t, \tag{11.14}$$

of which (11.13) is a rather special case. This arises quite naturally from the forced oscillator problem of Fig. 5.2, with $m = 1$, if the spring behaves the same way in compression as it does in extension, so that $F(-x) = -F(x)$, i.e. $F(x)$ is an odd function of x. If we expand $F(x)$ in a Taylor series about $x = 0$ to *two* terms instead of just one (cf. (5.9)) we then obtain $F(x) \doteqdot \alpha x + \beta x^3$, because there can be no x^2 term. The coefficient $\beta = \frac{1}{6}F'''(0)$ may be positive or negative, depending on the nature of the spring. We will take $\beta > 0$ in what follows, as in Fig. 5.2(b), and we will confine attention to steady oscillations of the system (11.14) at the forcing frequency Ω.

When the forcing amplitude A is small, so that the response x is also quite small, the cubic term in (11.14) is more or less negligible, and the system is essentially linear. The amplitude of the resulting steady oscillation is then greatest when Ω is close to the natural frequency $\omega = \sqrt{\alpha}$, though small damping ensures that the amplitude is finite even when $\Omega = \omega$ (Fig. 11.15(i), cf. Fig. 5.7(b)).

Fig. 11.15 *Sketch of the oscillation amplitude versus the forcing frequency Ω in (11.14), near the linear resonance point $\Omega = \omega$, for three different values of A.*

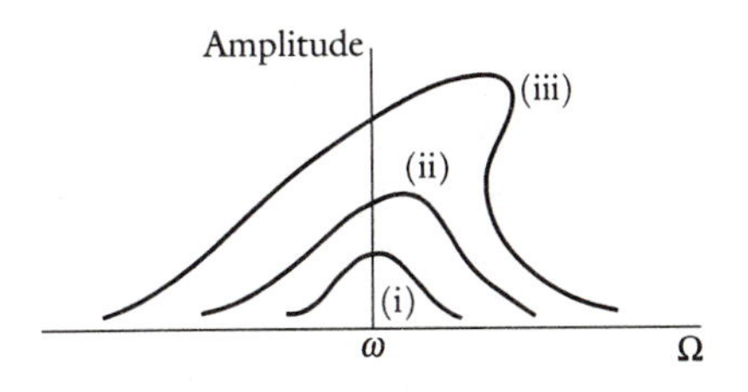

At somewhat larger forcing amplitudes A the cubic term in (11.14) becomes significant, and the response curve develops a pronounced asymmetry about $\Omega = \omega$, with the strongest response shifting to a higher forcing frequency Ω (Fig. 11.15(ii)). For larger forcing still, the response curve can actually 'turn over' (Fig. 11.15(iii)), and there is then a range of forcing frequencies Ω for which *three* different oscillations—all with frequency Ω—are possible.

There is a clear parallel here with the multiple *equilibria* of Chapter 10, and with Figs. 10.12 and 10.13 in particular. As one might guess, the oscillation with intermediate amplitude turns out to be unstable. Moreover, if we take the system and then gradually vary the forcing frequency Ω, sudden *jumps* in the amplitude of the resulting oscillation can occur, with associated *hysteresis* (Fig. 11.16, cf. Section 10.4). All this may be confirmed quite convincingly with the program NVARY.

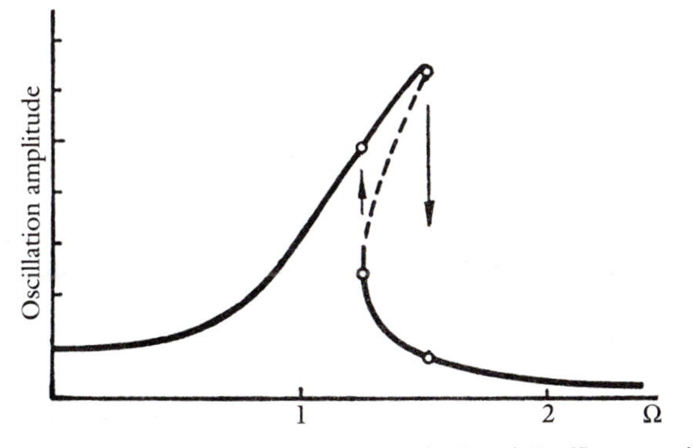

Fig. 11.16 The 'jump' phenomenon and hysteresis for the forced Duffing equation (11.14), with $k=0.1$, $\alpha=1$, $\beta=0.04$, $A=1$.

Exercises

11.1 *The van der Pol equation.* Use the program NXT to confirm the existence of the self-excited oscillations in Figs 11.4 and 11.6, and the way in which these oscillations are insensitive to the initial conditions. Then investigate how the oscillation period depends on ε when ε is large.

[While (11.2) cannot be solved exactly, so-called *asymptotic methods*, beyond the scope of this book, can be used to construct approximate solutions when ε is large, and in that limit the oscillation period is found to be $(3 - 2\log 2)\varepsilon \doteqdot 1.614\varepsilon$.]

11.2 *A 'simple' limit cycle*. As an example of eqn (11.4), consider the system

$$\dot{x} = y + \varepsilon x(1 - x^2 - y^2),$$
$$\dot{y} = -x + \varepsilon y(1 - x^2 - y^2),$$

where ε is a positive constant. *Prove* that it has a limit cycle, by changing to the variables r, θ, where $x = r\cos\theta$ and $y = r\sin\theta$. Then confirm this result with some numerical integrations in the x, y plane using the program `2PHASE`.

11.3 *Conditions for chaos*. Two *necessary* conditions for chaos are (i) the system must be nonlinear, and (ii) its phase space must be at least three-dimensional (see Section 11.3).

Can chaos occur in a *one*-dimensional *non*-autonomous system

$$\dot{x} = f(x, t)?$$

It is important to note that conditions (i) and (ii) are not *sufficient* for chaos, as we may see by considering the Euler equations (10.25) which govern the angular velocity components of a spinning rigid object:

$$A\dot{x} = (B - C)yz,$$
$$B\dot{y} = (C - A)zx,$$
$$C\dot{z} = (A - B)xy,$$

A, B, and C being constants. While this autonomous system is certainly nonlinear and three-dimensional, show that it is possible to deduce from these equations a *first integral*, i.e. a relationship

$$F(x, y, z) = \text{constant}.$$

It then follows immediately that chaotic motion cannot occur in this system. Why?

11.4 *The Lorenz and Rössler equations*. Use the program `NSENSIT` on p. 218 to confirm sensitive dependence on initial conditions for the Lorenz equations, with $r = 28$. Then use the program `NVARY` to confirm the period-doubling sequence for the Rössler equations shown in Fig. 11.13. What happens as c is gradually increased further to, say, $c = 6$?

11.5 *The forced cubic oscillator*. Use the program NXT to confirm the chaotic oscillations and sensitivity to initial conditions shown in Fig. 11.1 for the equation

$$\ddot{x} + k\dot{x} + x^3 = A\cos\Omega t.$$

Then use the program NXTWAIT, on p. 221, to show that the same system has five coexisting limit cycles when $k = 0.08$, $\Omega = 1$, $A = 0.2$. (Try initial values for x of 1, 0.2, -0.9, -0.7 and 0, with $\dot{x} = 0$ at $t = 0$.)

11.6 *Chaos from a simple iteration*. The period-doubling route to chaos can be seen very clearly in the iterative equation

$$x_{n+1} = \lambda x_n(1 - x_n), \tag{11.15}$$

which has its roots in simple population models (cf. (3.15)).

Write a program to generate $x_1, x_2, x_3 \ldots$ from a starting value x_0 in the range $0 < x_0 < 1$. Confirm that:

(i) $x_n \to 0$ when $0 < \lambda < 1$;
(ii) $x_n \to 1 - 1/\lambda$ when $1 < \lambda < 3$;
(iii) x_n settles down to an oscillation between two different values when $3 < \lambda < 3.449$;
(iv) x_n settles down to an oscillation between *four* different values for slightly greater values of λ; and
(v) this period-doubling continues (in fact, indefinitely) as λ approaches 3.570, beyond which there is a range of λ for which the eventual behaviour of x_n with n is chaotic (Fig. 11.17).

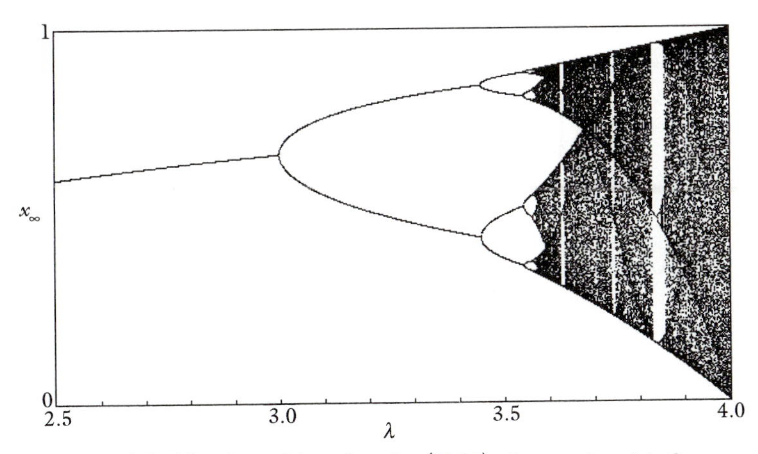

Fig. 11.17 A period-doubling 'cascade' to chaos for (11.15). At any given $\lambda > 3$, x_n eventually settles into an oscillation involving the values shown.

12 The not-so-simple pendulum

12.1 Introduction

The pendulum is one of the oldest subjects of scientific inquiry, yet it is still capable of springing surprises.

One of its most remarkable properties was discovered in 1908 by Andrew Stephenson, a mathematics lecturer at Manchester University. He showed that it is possible to maintain a rigid pendulum stably in its 'upside-down' position by making its pivot vibrate up and down at high frequency (Fig. 12.1).

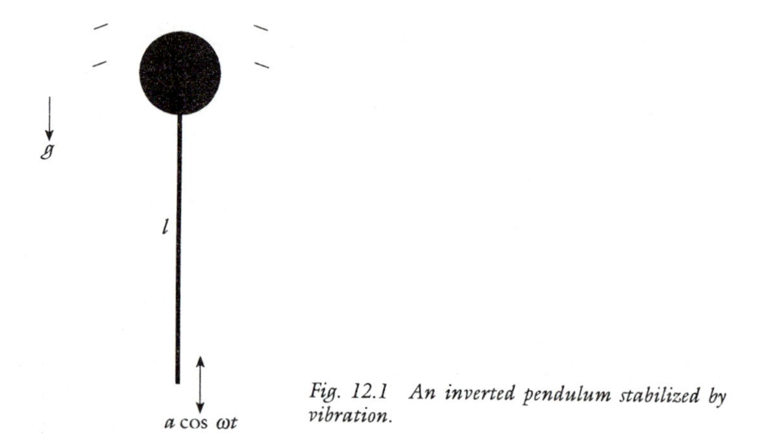

Fig. 12.1 *An inverted pendulum stabilized by vibration.*

Suppose, for simplicity, that the pendulum is a light rod of length l with a point mass at one end. When the pivot vibrations have an amplitude a which is small compared with l, Stephenson showed that the inverted state will be stable if

$$\omega > \frac{\sqrt{2gl}}{a}, \tag{12.1}$$

where ω denotes the vibration frequency of the pivot.

In practice ω needs to be quite large; with $l = 10\,\text{cm}$ and $a = 1\,\text{cm}$, say, the minimum value of $\omega/2\pi$ turns out to be about 22 Hz. Even so, it is still not at all easy to visualize how the stability of the inverted state is achieved. We are all familiar, perhaps, with the idea of balancing an upright pole on the palm of one hand, but that involves continually moving the 'pivot' from side to side in response to how the pole is toppling over at the time. In contrast, there is no such 'feedback' in Fig. 12.1, and the oscillations of the pivot are strictly up-and-down.

Subsequent studies have shown that the inverted state can be stabilized with lower drive frequencies ω and larger drive amplitudes a, provided these lie within the shaded region in Fig. 12.2, i.e. between the curves L and R. The first of these corresponds to Stephenson's criterion (12.1) when a/l is small.

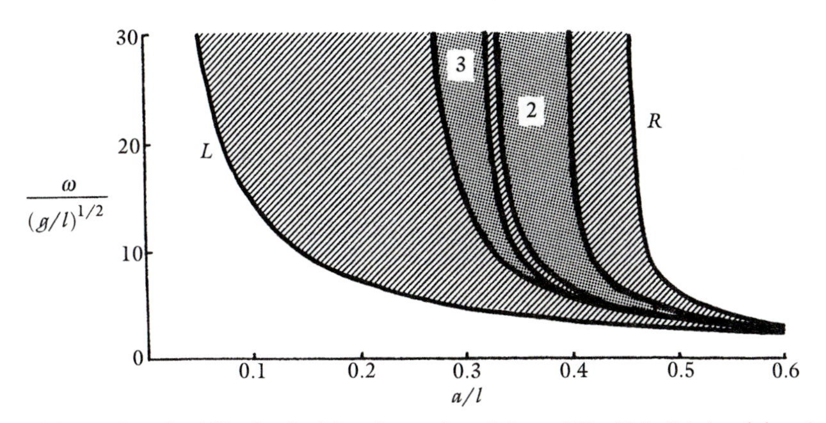

Fig. 12.2 Region of stability for the driven inverted pendulum of Fig. 12.1. Frictional damping is incorporated in the model, with $\bar{k}=0.1$ (see eq. (12.13)). Upside-down 'dancing' provides an alternative to the classical inverted state in regions 2 and 3 (see Fig. 12.3).

But even this is not the full story. Only a year or two ago I discovered that in regions 2 and 3 of Fig. 12.2 the pendulum has *two different ways* in which it can avoid falling over. The first is Stephenson's, in which it gradually wobbles closer and closer to the upward vertical as time goes on, but the other has the pendulum 'dancing' persistently about the upward vertical, with its bob at all times higher than the pivot (Fig. 12.3). In region 2 the inverted pendulum bobs twice in succession on one side of the upward vertical before being flung across to the other side (Fig. 12.3(a)), while in region 3 there are three successive bobs on one side of the upward vertical before the second half of each oscillation cycle (Fig. 12.3(b)). These upside-down limit cycles only take the place of the simple state in Fig. 12.1 if the pendulum is given a sufficient nudge away from the upward vertical at $t = 0$ (Ex. 12.1).

We shall present an even more counter-intuitive result on inverted pendulums in Section 12.5, but first we consider some wider aspects of pendulum motion which are significant for modern research.

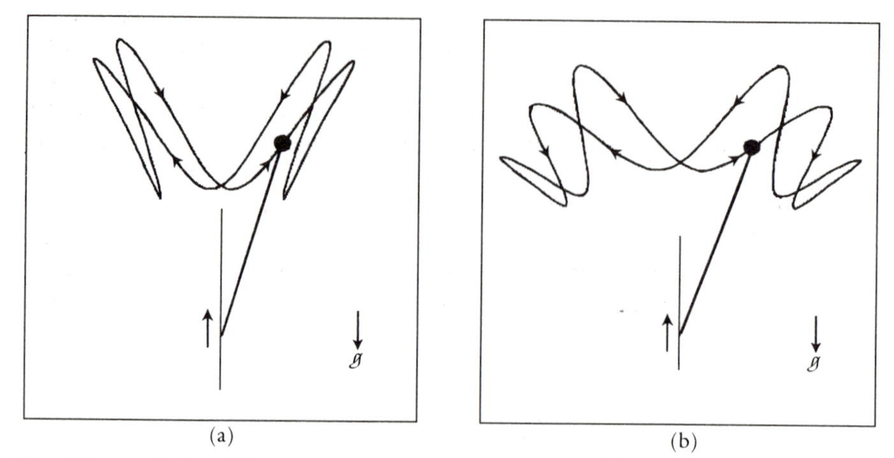

Fig. 12.3 *Upside-down 'dancing' of a driven inverted pendulum. (a) An oscillation with frequency $\frac{1}{4}\omega$ from region 2 in Fig. 12.2, with $a/l = .45$ and $\omega = 5(g/l)^{1/2}$. It may be obtained with, say, $\theta = \pi$ and $\mathrm{d}\theta/\mathrm{d}t = 1.3(g/l)^{1/2}$ at $t = 0$. (b) An oscillation with frequency $\frac{1}{6}\omega$ from region 3, with $a/l = .33$ and $\omega = 10(g/l)^{1/2}$. It may be obtained with, say, $\theta = \pi$ and $\mathrm{d}\theta/\mathrm{d}t = 2.6(g/l)^{1/2}$ at $t = 0$.*

We begin by turning the clock back to 1687, and to some pendulums which were, at the time, having a deep influence on the evolution of dynamics itself.

12.2 Pendulums from the past

In the early days of dynamics, problems involving the collision of two bodies were among the first to be tackled successfully, and they played a major part in the development of the subject as a whole. Indeed, Newton's third law 'action and reaction are equal and opposite' arose largely from his study of such problems.

At that time, the chief experimental difficulty lay in the accurate measurement of the velocities of the two bodies immediately before and after the

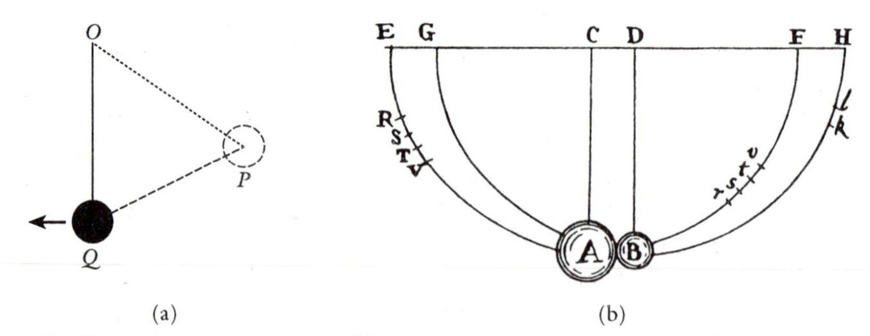

Fig. 12.4 *(a) A simple pendulum. (b) A collision experiment from Newton's* Principia *(1687).*

collision itself, and it was here that pendulums came to the rescue. Cunning use was made of what Newton called 'a proposition well known to Geometers', namely that when a pendulum bob is drawn aside to a point P and released from rest it acquires at its lowest point Q a speed which is *proportional to the distance QP* (see Fig. 12.4(a) and Ex. 12.2). The idea, then, was to conduct collision experiments using two pendulum bobs (Fig. 12.4(b)), so that the problem of measuring the instantaneous velocities reduced simply to the measurement of a few distances.

This was not the only way in which pendulums helped shape the course of dynamics at the time. In the records of the meeting of the Royal Society on 16th May 1666 we find

> It being mentioned by Mr. Hooke, that the motion of the celestial bodies might be represented by pendulums, it was order'd, that this should be shewed at the next meeting.

This 'representation' turned out, in fact, to be no more than a loose analogy (Fig. 12.5), but there is evidence that it helped steer contemporary thinking towards the idea that the planets might be subject to a continual force of attraction towards the Sun.

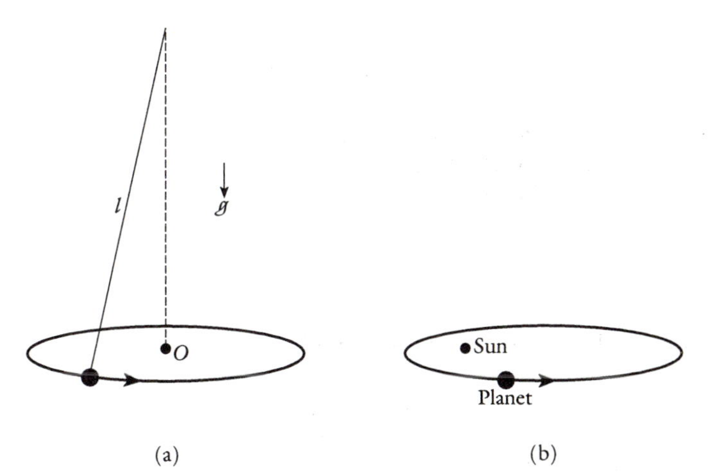

Fig. 12.5 Robert Hooke's loose analogy between (a) three-dimensional pendulum motion, which takes place in an ellipse centred *on O if the amplitude is small, and (b) planetary orbits, which are elliptical with one* focus *at the Sun.*

For present purposes, however, one of the most interesting pendulums emerged rather later, in work by Euler and Daniel Bernoulli published in the 1730s. This **double pendulum** consists of one pendulum suspended freely from another, but with both constrained to swing in the same vertical plane (Fig. 12.6). If each is a light rod of length l, with point masses m_1 and m_2, the

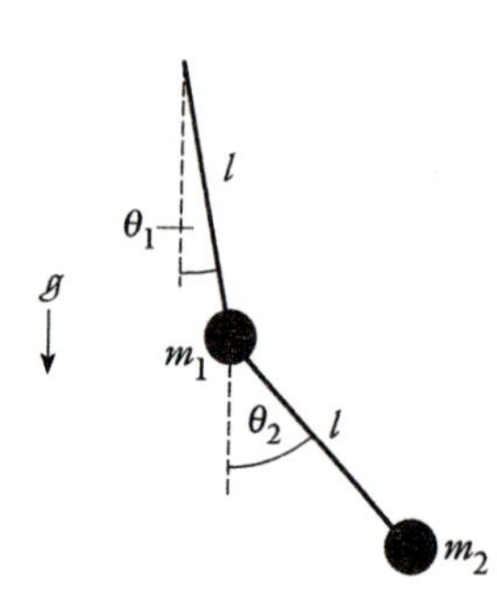

Fig. 12.6 A double pendulum.

equations for *small-amplitude* motion about the downward vertical turn out to be

$$\begin{aligned}\frac{d^2\theta_1}{dt^2} + m\frac{d^2\theta_2}{dt^2} + \frac{g}{l}\theta_1 = 0,\\ \frac{d^2\theta_2}{dt^2} + \frac{d^2\theta_1}{dt^2} + \frac{g}{l}\theta_2 = 0,\end{aligned} \tag{12.2a, b}$$

where

$$m = \frac{m_2}{m_1 + m_2}. \tag{12.3}$$

It is then a relatively straightforward task to substitute $\theta_1 = A\cos\omega t$, $\theta_2 = B\cos\omega t$ into these equations and find that the system has *two* natural frequencies of oscillation ω given by

$$\omega^2 = \frac{g/l}{1 \pm \sqrt{m}}, \tag{12.4}$$

the corresponding pendulum motions being such that

$$\frac{\theta_2}{\theta_1} = \pm\frac{1}{\sqrt{m}} \tag{12.5}$$

(cf. (5.24), (5.25)). In the low-frequency mode, obtained by taking the upper sign in these expressions, the two pendulums swing in the same direction at any given moment, while in the high-frequency mode they swing in opposite directions.

In the case when the two masses are equal ($m = \frac{1}{2}$) the frequency of the fast mode is about $2\frac{1}{2}$ times that of the slow mode (see Fig. 5.10), but if m_2/m_1 is large then m is almost equal to 1, and the two natural frequencies are much more widely separated. In this limit, the slow mode has θ_1 and θ_2 almost equal, and the pendulums swing like a single pendulum of length $2l$.

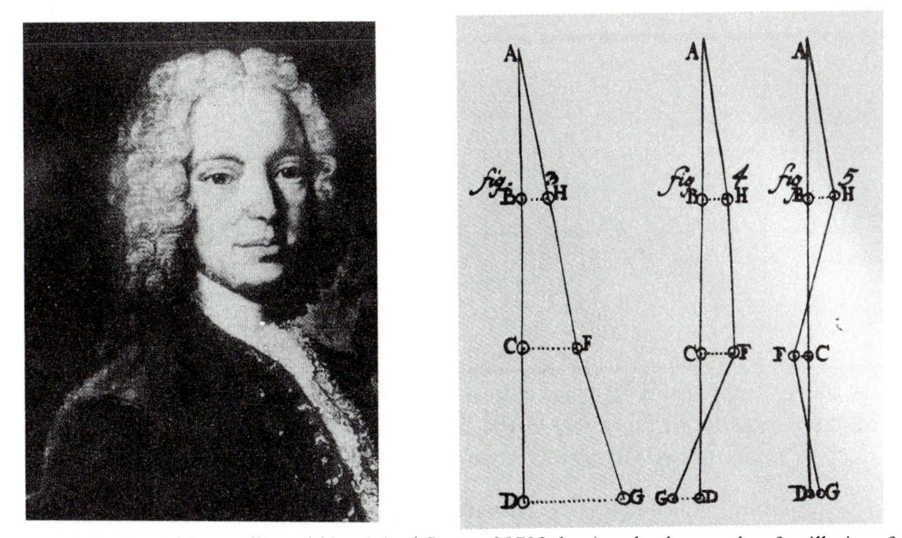

Fig. 12.7 Daniel Bernoulli, and his original figure of 1738 showing the three modes of oscillation of a triple pendulum.

The fast mode, on the other hand, then has θ_1 and θ_2 almost equal and opposite, so that the upper mass oscillates to and fro while the lower, much greater mass remains almost stationary.

In a similar way, if N pendulums hang down, one from another, there are N different natural frequencies of small oscillation. In the case $N = 3$ these turn out to be

$$\omega_1 = 0.64\left(\frac{g}{l}\right)^{1/2}, \qquad \omega_2 = 1.51\left(\frac{g}{l}\right)^{1/2}, \qquad \omega_3 = 2.51\left(\frac{g}{l}\right)^{1/2}, \tag{12.6}$$

in the special case when the pendulums are light rods of equal length l with equal point masses at the ends (Fig. 12.7).

12.3 A vibrated pendulum

We saw in Section 12.1 how even a single pendulum can behave in a rather curious fashion *if its pivot is made to vibrate up and down.*

It is, in fact, remarkably easy to obtain the equation of motion for such a pendulum. According to the principles of mechanics we have only to take the 'usual' equation, (5.2):

$$\frac{d^2\theta}{dt^2} + \frac{g}{l}\sin\theta = 0, \tag{12.7}$$

and replace g, the acceleration due to gravity, by $g - d^2h/dt^2$, where d^2h/dt^2 denotes the downward acceleration of the pivot. (This is why we feel

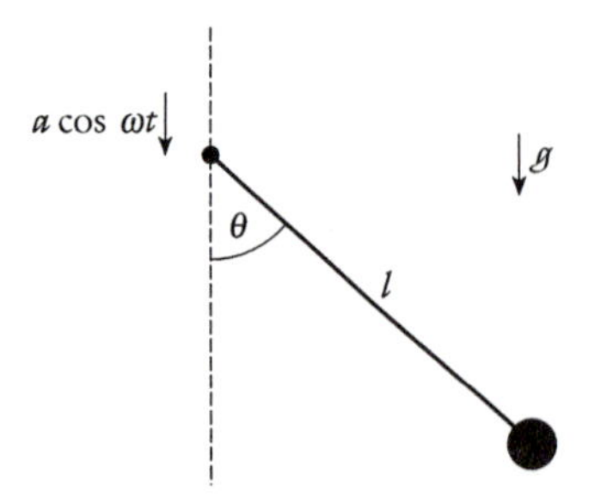

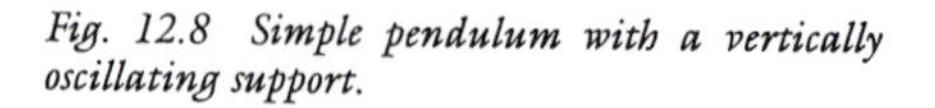
Fig. 12.8 Simple pendulum with a vertically oscillating support.

less heavy than usual if we are in a lift which is accelerating downwards; 'effective' gravity, so far as we in the lift are concerned, is only $g - \mathrm{d}^2h/\mathrm{d}t^2$.)

Let us make the pivot vibrate, then, so that at time t it is a distance

$$h = a\cos\omega t \tag{12.8}$$

measured downwards from some fixed point. Then $\mathrm{d}^2h/\mathrm{d}t^2 = -a\omega^2\cos\omega t$, so the equation of motion is

$$\frac{\mathrm{d}^2\theta}{\mathrm{d}t^2} + k\frac{\mathrm{d}\theta}{\mathrm{d}t} + \left(\frac{g}{l} + \frac{a\omega^2}{l}\cos\omega t\right)\sin\theta = 0, \tag{12.9}$$

where we have included a simple linear damping term with constant k, simulating the effects of friction.

If we change to the dimensionless time variable

$$\tilde{t} = t/(l/g)^{1/2} \tag{12.10}$$

we may rewrite (12.9) as

$$\ddot{\theta} + \tilde{k}\dot{\theta} + (1 + \tilde{a}\tilde{\omega}^2\cos\tilde{\omega}\tilde{t})\sin\theta = 0, \tag{12.11}$$

where a dot denotes differentiation with respect to $\tilde{t}$. There are now just three (dimensionless) parameters involved, namely

$$\tilde{a} = \frac{a}{l}, \qquad \tilde{\omega} = \frac{\omega}{(g/l)^{1/2}}, \qquad \tilde{k} = \frac{k}{(g/l)^{1/2}}. \tag{12.12}$$

These act as convenient measures of the amplitude and frequency of the pivot motion, and the frictional damping of the pendulum itself.

We may, in turn, recast (12.11) as an autonomous first-order system:

$$\begin{aligned} \dot{\theta} &= y, \\ \dot{y} &= -\tilde{k}y - (1 + \tilde{a}\tilde{\omega}^2\cos\tilde{\omega}\tilde{t})\sin\theta, \\ \dot{\tilde{t}} &= 1 \end{aligned} \tag{12.13}$$

and the program VIBRAPEN on p. 228 integrates this by the Runge–Kutta method and displays the outcome as a simple animation.

An unusual instability

Note first that the pendulum in Fig. 12.8 does not *have* to swing at all; if $\theta = \dot{\theta} = 0$ at $t = 0$ then the solution to (12.11) satisfying those initial conditions is simply $\theta = 0$ for all time t.

But this purely up-and-down motion of the pendulum is not always stable. Suppose that $|\theta|$ is small, so that we may linearize (12.11) by replacing $\sin\theta$ by θ. Then, in the absence of damping, the further change of variable $\tilde{t} = T/\tilde{\omega}$ leads to **Mathieu's equation**

$$\frac{d^2\theta}{dT^2} + (\alpha + \beta\cos T)\theta = 0, \tag{12.14}$$

where $\alpha = 1/\tilde{\omega}^2$ and $\beta = \tilde{a}$. This equation has been studied in many different physical contexts, and the solution $\theta = 0$ is known to be unstable if the constants α_1, β lie in one of the shaded regions in Fig. 12.9. Notice, in particular, that if β is small there is a small instability region centred on $\alpha = \frac{1}{4}$, i.e. $\tilde{\omega} = 2$.

Suppose, then, that the pivot in Fig. 12.8 vibrates up and down with fairly small amplitude but with a frequency which is *twice* the natural frequency of the pendulum itself, so that $\tilde{\omega} = 2$. The $\theta = 0$ state is then unstable, and an initially small disturbance will lead to a swinging motion of steadily increasing amplitude.

In practice, a small amount of frictional damping is inevitable, and this slightly erodes the sharp tips of the instability regions in Fig. 12.9, so that even if $\tilde{\omega}$ is exactly 2 the instability only occurs if $\tilde{a}$ exceeds some critical

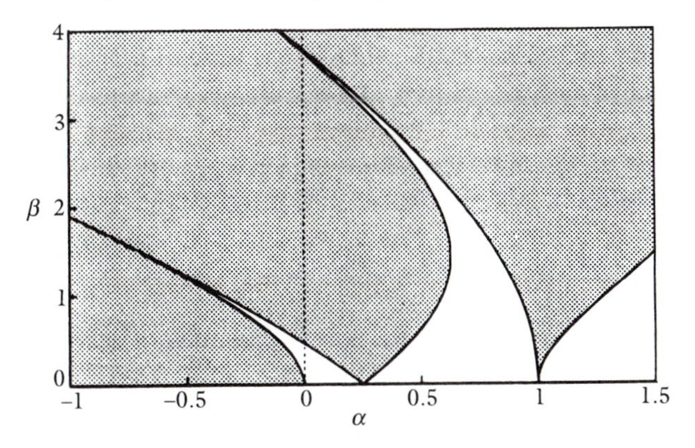

Fig. 12.9 Stability diagram for Mathieu's equation (12.14). The $\theta = 0$ state is unstable in the shaded regions.

value. Moreover, as the amplitude of the swinging motion increases, the linearization $\sin\theta \doteqdot \theta$ ceases to be valid, and we have to turn to numerical integrations of (12.11) using, say, VIBRAPEN. These show that if $\tilde{a}$ is not too large the pendulum settles eventually into a limit cycle oscillation about the downward vertical with frequency $\frac{1}{2}\omega$ (Ex. 12.3).

12.4 Chaotic pendulums

Pendulums can provide some of the most vivid practical demonstrations of chaotic motion, and one or two are even commercially available as so-called 'executive toys'.

A good example is the 'pendumonium',* a four-pendulum version of matchstick man devised by Professor N. Rott of Stanford University (Fig. 12.10). The head, shoulders and trunk are joined rigidly and pivoted at the

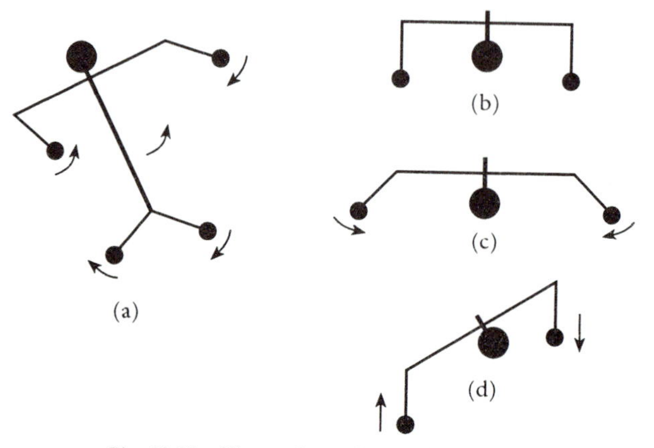

Fig. 12.10 The pendumonium: a chaotic toy.

neck, but there are two additional pendulums as arms and a further pendulum in the form of a pair of legs. Even with the trunk and legs removed it is still a three-pendulum system (Fig. 12.10(b)), well capable of chaotic motion when spun about its pivot, with the arms, in particular, whirling this way and that in a most erratic and eye-catching manner. The system is also cunningly designed so that the swinging mode of oscillation in Fig. 12.10(c) has exactly twice the frequency of the up-and-down mode in Fig. 12.10(d); one type of oscillation can then gradually change into the other, in part by the instability mechanism which we have just described in Section 12.3.

* Made by Hands On Instruments, PO Box 52044, Palo Alto, California 94303, USA.

A chaotic double pendulum

Consider now the classical double pendulum of Fig. 12.6, and recall that the equations (12.2a, b) are for small-amplitude motion only. The full equations of motion for the system can be shown to be

$$\frac{d^2\theta_1}{dt^2} + m\frac{d^2\theta_2}{dt^2}\cos(\theta_2 - \theta_1) - m\left(\frac{d\theta_2}{dt}\right)^2 \sin(\theta_2 - \theta_1) + \frac{g}{l}\sin\theta_1 = 0,$$

$$\frac{d^2\theta_2}{dt^2} + \frac{d^2\theta_1}{dt^2}\cos(\theta_2 - \theta_1) + \left(\frac{d\theta_1}{dt}\right)^2 \sin(\theta_2 - \theta_1) + \frac{g}{l}\sin\theta_2 = 0, \tag{12.15a,b}$$

in the absence of friction. As a check, note that we recover (12.2a, b) if we assume that θ_1, θ_2 and their various derivatives are small and then linearize the equations.

Let us now liven up this system by vibrating the pivot up and down, as in Fig. 12.8, so that g is replaced in (12.15a, b) by $g + a\omega^2 \cos \omega t$. In contrast to the single-pendulum case (12.13), the resulting first-order autonomous system consists of *five* equations, and we integrate these numerically with the program PENDOUBL.

Figure 12.11 shows an example of chaotic motion when $m = 0.1$, so that m_2 is much smaller than m_1. Here $\tilde{a} = 0.35$, $\tilde{\omega} = 2$ and $\tilde{k} = 0.1$, so frictional damping has been included in the equations (cf. (12.12)). While the motion certainly *looks* chaotic enough, the real test—as explained in Chapter 11—lies in how sensitive the motion is to the initial conditions.

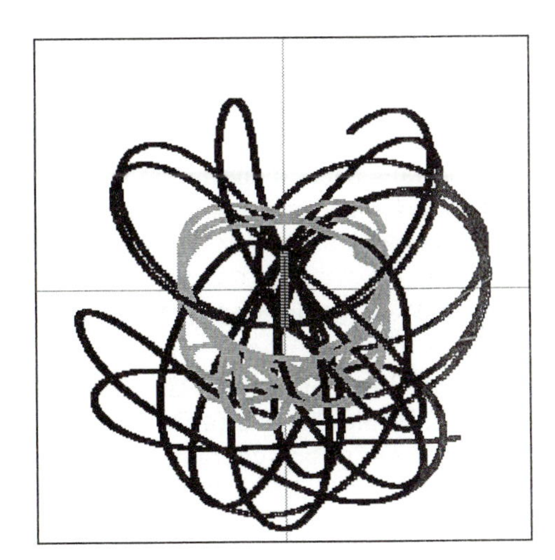

Fig. 12.11 Typical chaotic bob trajectories for the double pendulum of Fig. 12.6, when its pivot is vibrated up and down.

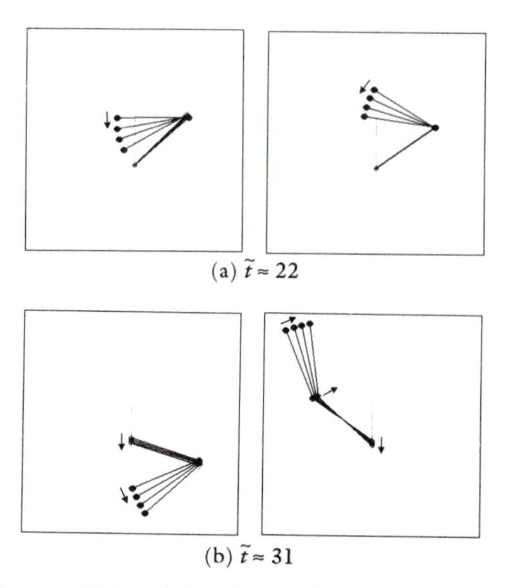

Fig. 12.12 Sensitivity to initial conditions for a vibrated and chaotic double pendulum, with $\tilde{a}=0.35$, $\tilde{\omega}=2$ *and* $\tilde{k}=0.1$. *The first motion sequence in each case arises from the initial conditions* $\theta_1=\theta_2=1$, $\dot{\theta}_1=\dot{\theta}_2=0$, *while the second arises from* $\theta_1=\theta_2=1.005$, $\dot{\theta}_1=\dot{\theta}_2=0$ *at* $t=0$.

This sensitivity is evident from Fig. 12.12, where a change in those conditions of just 1 part in 200 is seen leading to a significantly different outcome after only seven downstrokes of the pivot (Fig. 12.12(a)). Moreover, after just ten downstrokes the two solutions have so diverged that the second, lighter pendulum is actually whirling in opposite directions in the two cases (Fig. 12.12(b)).

If we run PENDOUBL with the two sets of initial values in Fig. 12.12 we obtain some insight into the source of this divergence. For $0 < \tilde{t} < 12$ the lighter pendulum m_2 is flung three times round its pivot, anticlockwise, and it then has a relatively 'quiet' period for $12 < \tilde{t} < 21$. In the interval $21 < \tilde{t} < 25$ it makes another two anticlockwise rotations and *just* manages to overshoot its vertically upright position for a third if $\theta_1 = \theta_2 = 1$ initially but fails to do this, and falls back, if the initial values are $\theta_1 = \theta_2 = 1.005$. Not surprisingly, the two motions develop quite differently thereafter.

12.5 Not quite the Indian Rope Trick

We began this chapter with the inverted pendulum of Fig. 12.1, stabilized by vibration. While this is a fairly well-known curiosity of classical mechanics, it does not seem to be generally known that the same 'trick' can be performed with *any finite number* of linked pendulums, *all balanced on top of one another*.

This, at least, is one implication of the following theorem, which I proved just a few years ago.

An inverted pendulum theorem (1993)

Suppose N pendulums hang down, one from another, under gravity g, the uppermost being attached to a pivot. Let ω_{min} denote the smallest of the natural frequencies of oscillation of this system, and let ω_{max} denote the largest. Suppose, too that ω_{max}^2 is much larger than ω_{min}^2 (which is usually the case when N is 2 or more).

Then *the whole system can be stabilized in its upside-down state by vibrating the pivot up and down with amplitude a and frequency ω such that*

$$a < \frac{0.450g}{\omega_{max}^2} \qquad (12.16a)$$

and

$$a\omega > \frac{\sqrt{2}g}{\omega_{min}}. \qquad (12.16b)$$

The stable region in the a, ω plane therefore has the characteristic shape indicated in Fig. 12.13, where the straight line BC corresponds to (12.16a) and the curve AB corresponds to (12.16b).

The sketches indicate what happens if we gradually leave the stable region. If the frequency ω is reduced so as to violate (12.16b), the pendulums collapse by slowly wobbling down on one side of the vertical. If, instead, the drive amplitude a is increased so as to violate (12.16a), the collapse is more

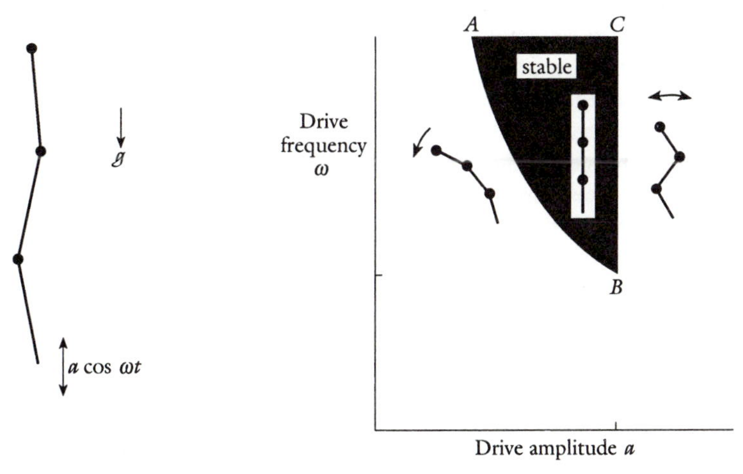

Fig. 12.13 An inverted pendulum theorem.

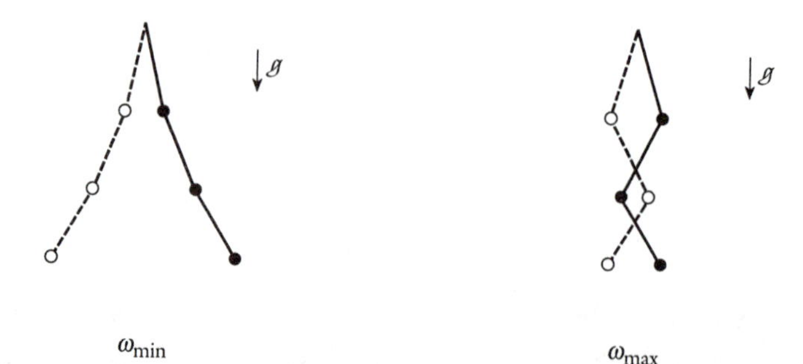

Fig. 12.14 The undriven, downward-hanging triple pendulum; typical motions corresponding to ω_{min} *and* ω_{max}.

dramatic, via a buckling oscillation of steadily increasing amplitude.

Without attempting a proof of the theorem here, we note in passing that it makes use of Mathieu's equation (12.14) and, in particular, the tiny region of stability in Fig. 12.9 for *negative* α. Indeed, the mysterious factor 0.450 in (12.16a) corresponds to where the upper boundary of that region meets the β-axis.

Remarkably, then, the two numbers ω_{min} and ω_{max} are all that we need to know about the particular system in question in order to apply (12.16). The theorem works by relating the stability of the vibrated, inverted state to just these two simple oscillation properties of the system when it is in its *non*-inverted state with its pivot held *fixed* (Fig. 12.14).

The numerical values of ω_{min} and ω_{max} depend, of course, on the number of pendulums involved, and their various shapes and sizes and mass distributions. Thus for the particular triple pendulum in Fig. 12.7 we would have $\omega_{min} = 0.64(g/l)^{1/2}$ and $\omega_{max} = 2.51(g/l)^{1/2}$, according to (12.6).

If many pendulums are involved, ω_{max} is typically quite large, so the pivot amplitude a has to be small in order to satisfy (12.16a). This in turn means that the pivot *frequency* ω has to be very *large*, in order to satisfy (12.16b), and in this way it becomes comparatively difficult to stabilize a long chain of pendulums in its inverted state, as one might well expect.

But do the strange predictions of the theorem really work out in practice? In order to test this, one of my colleagues, Tom Mullin, conducted some remarkable experiments.

Figure 12.15 shows an example, with three linked rods, each of length $l = 19$ cm, one being attached to a pivot (Fig. 12.15(a)). With a pivot vibration amplitude a of about 1 cm and a pivot frequency $f = \omega/2\pi$ of 40 Hz or so this triple pendulum can indeed be stabilized in its inverted state (Fig. 12.15(b)).

The rods in Fig. 12.15 are joined by two low-friction bearings, and each of these weighs nearly twice as much as one of the rods. As a result, $\omega_{min} \doteqdot 0.729(g/l)^{1/2}$ and $\omega_{max} \doteqdot 2.174(g/l)^{1/2}$ for this particular system, so the

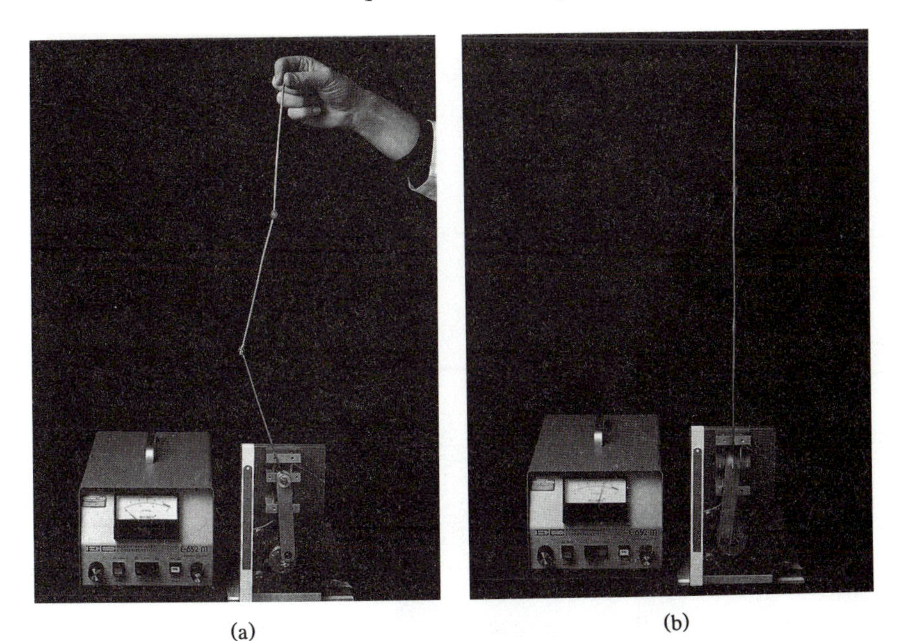

(a) (b)

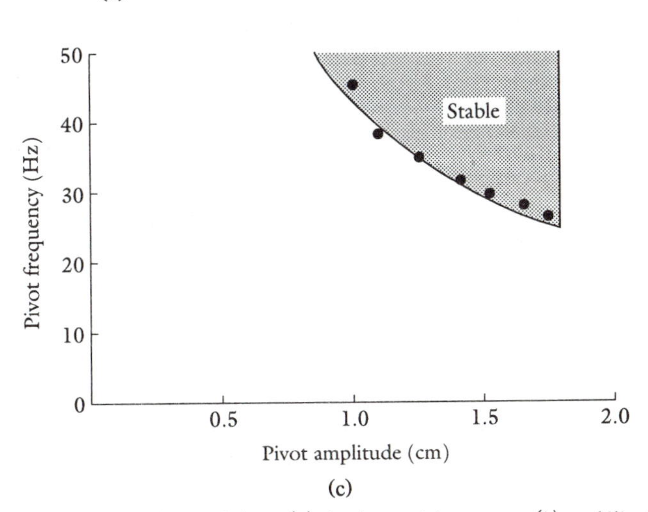

(c)

Fig. 12.15 An inverted triple pendulum (a) in its undriven state (b) stabilized by vertical vibrations of its pivot. (c) Comparison between theory and experiment.

theorem (12.16) predicts stability of the inverted state if

$$\frac{a}{l} < 0.095 \tag{12.17a}$$

and

$$\frac{a}{l} \cdot \frac{\omega}{(g/l)^{1/2}} > 1.94, \tag{12.17b}$$

Fig. 12.16 An inverted triple pendulum recovering from a severe initial disturbance; $a \approx 1.4$ cm, $f \approx 35$ Hz.

giving the region of stability shown in Fig. 12.15(c). This was tested experimentally, for various different values of the drive amplitude a, by first stabilizing the inverted state and then gradually reducing the drive frequency $f = \omega/2\pi$ until the system collapsed (at the points marked •). The predictions of the inverted pendulum theorem are therefore in good agreement with the experimental results.

We were particularly surprised by just *how* stable the inverted triple pendulum could be in the actual experiment. The theorem itself is based on linear stability theory, so guarantees stability only with respect to infinitesimally small disturbances. Yet in the experiments we found that we could gently push the tower of pendulums over by as much as 45° or so, and provided we kept it reasonably straight in the process it would then gradually wobble back to the upward vertical (Fig. 12.16).

It is just possible, in fact, that some readers may have seen these experiments, because they featured in October 1995 on BBC TV's long-running science programme *Tomorrow's World*. A loose analogy was drawn there with the famous but elusive 'Indian Rope Trick', in which a length of rope is thrown into the air and stays upright, defying gravity (Fig. 12.17). And we did, indeed, show on the programme how we could perform part of that trick.

Fig. 12.17 'Karachi' and his son 'Khydar'. They claimed to have performed the Indian Rope Trick in 1935, but neither this claim, nor any others, has ever been substantiated.

Our 'rope' is in practice a length of very floppy PVC-coated net-curtain wire (cf. Fig. 10.8). We had thought at first that stabilizing it by vibration would be impossible, for if the wire is viewed as an infinite number of infinitely-short pendulums then ω_{max} is infinite, so (12.16a) cannot be satisfied. But this kind of argument clearly relates to a *perfectly flexible* piece of wire or rope, whereas ours, like any other, has *some* resistance to bending, so the theorem simply does not apply.

In any event, the 'trick' can certainly be seen in Fig. 12.18. The vibration frequency of the base is rather lower than those used in the inverted triple pendulum experiments, but crucial nonetheless: without it the wire is long enough to be unstable to even the smallest disturbance in its upright position, and flops right down instead (Fig. 12.18(a)).

It should be noted in conclusion that we are claiming to perform—in our own peculiar way—only *part* of the most elusive magic trick in history. There should, in truth, be a grand finale, in which a small boy climbs up the system and disappears at the top, but we have no plans to attempt this.

Exercises

12.1 *The inverted pendulum.* Use the program VIBRAPEN on p. 228, together with Fig. 12.2, to confirm that the inverted state really can be stabilized. Confirm, too, the existence of the upside-down dancing oscillations in Fig. 12.3.

12.2 *Newton's collision experiments.* Use conservation of energy to prove the 'proposition well known to Geometers' which Newton used in connection with the pendulum collision experiment of Fig. 12.4.

12.3 *An unusual instability.* Use VIBRAPEN to confirm the instability of the downward-hanging state of the pendulum in Fig. 12.8 when $\tilde{a} = 0.1$ and $\tilde{\omega} = 2$, with $\tilde{k} = 0.1$. Compare the different outcomes when, say, $\tilde{\omega} = 1.7$ or 2.3.

Then use the program PENDOUBL to investigate the corresponding instability of the *double* pendulum in Fig. 12.6 with, say, $m = 0.5$ and $\tilde{a} = a/l = 0.1$. Can *either* of the modes of oscillation (12.4) be excited by vibrating the pivot up and down at twice the appropriate natural frequency?

12.4 *Non-uniqueness and chaos.* Use PENDOUBL to investigate the double pendulum of Fig. 12.6 when its pivot is vibrated up and down in the manner of Fig. 12.8, taking $m = 0.1$ with $\tilde{\omega} = 2$ and $\tilde{k} = 0.1$.

Show that if $\tilde{a} = 0.1$ the system eventually settles to *either* the downward-hanging state *or* a regular oscillation in which the two pendulums swing in opposite directions at any given moment.

(a)

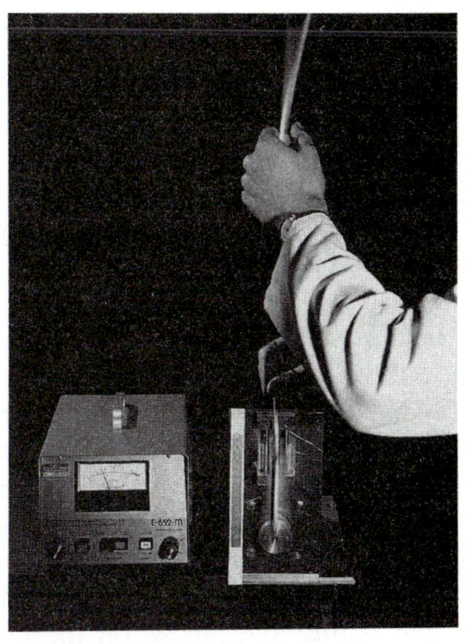

(b)

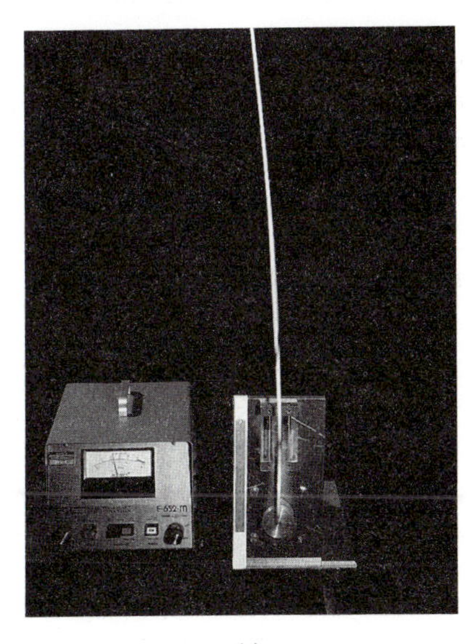

(c)

Fig. 12.18 A length of floppy wire, clamped at its lowest point, is stabilized in its 'upright' position by vibrating its base up and down. Without such vibrations the only stable position for the wire would be a completely 'flopped-down' state, as in (a) (and in Fig. 10.8(c)).

Show that if $\tilde{a} = 0.25$ the downward-hanging state is unstable, and the system now settles to *either* the regular swinging oscillation described above *or* a regular 'whirling' motion in which both pendulums continually rotate about the pivot.

Finally, take $\tilde{a} = 0.35$ and confirm the existence of chaotic whirling motions, and in particular the sensitivity to initial conditions claimed in Fig. 12.12.

12.5 *The inverted double pendulum.* Use PENDOUBL to investigate the inverted state when $m = 0.5$ and $\tilde{k} = 0.2$.

Confirm first that it is stable to small disturbances when, say, $\tilde{a} = 0.1$ and $\tilde{\omega} = 25$, and show that this is in keeping with the general theorem in Section 12.5. Investigate the extent to which the inverted state is stable to larger disturbances, trying, in particular, $\theta_1 = \theta_2 = 2.5$ at $t = 0$ and $\theta_1 = 3.3$, $\theta_2 = 2.9$ at $t = 0$, with $\dot{\theta}_1 = \dot{\theta}_2 = 0$ initially in each case.

Keep $\tilde{\omega} = 25$ and show that if $\tilde{a}$ is either too small or too large the inverted state is unstable. Show, on the other hand, that if $\tilde{a} = 0.11$, say, there is a second way in which the system can avoid collapse, namely by upside-down dancing oscillations akin to those in Fig. 12.3(a).

Further reading

Chapter 1: Introduction

For the history of mathematics, try

1. Hollingdale, S. (1989). *Makers of mathematics*. Penguin.
2. Stillwell, J. (1989). *Mathematics and its history*. Springer-Verlag.
3. Kline, M. (1972). *Mathematical thought from ancient to modern times*. Oxford University Press.
4. Grattan-Guinness, J. (ed.) (1994). *Companion encyclopaedia of the history and philosophy of the mathematical sciences*. Routledge.

See, in particular, part 8 of the book edited by Gratton-Guinness for articles on the history of mechanics.

Chapter 2: A brief review of calculus

One of the best introductions to calculus is Chapter 8 of

5. Courant, R. and Robbins, H. (1996). *What is mathematics*? (2nd edn). Oxford University Press.

Two semi-'popular' books are

6. Beckmann, P. (1993). *A history of* π. Barnes and Noble, New York.
7. Maor, E. (1994). *e, the story of a number*. Princeton University Press.

To take the calculus further, try

8. Finney, R. L. and Thomas, G. B. (1994). *Calculus*, (2nd edn). Addison Wesley.
9. Salas, S. L. and Hille, E. (1995). *Calculus*, (7th edn). Wiley.

An interesting recent book with a strongly historical slant is

10. Hairer, E. and Wanner, G. (1996). *Analysis by its history*. Springer-Verlag.

Chapter 3: Ordinary differential equations

One well-established text on applied mathematical methods is

11. Kreyszig, E. (1993). *Advanced engineering mathematics*, (7th edn). Wiley.

and Part A is concerned with ODEs.

Good books specifically on differential equations include

12. Simmons, G. F. (1991). *Differential equations, with applications and historical notes*, (2nd edn). McGraw-Hill.
13. Boyce, W. E. and Di Prima, R. C. (1997). *Elementary differential equations* (6th edn). Wiley.
14. Farlow, S. J. (1994). *An introduction to differential equations and their applications*. McGraw-Hill.
15. Birkhoff, G. and Rota, G. C. (1989). *Ordinary differential equations* (4th edn). Wiley.

For a more geometric view of the subject try

16. Hubbard, J. H. and West, B. H. (1991). *Differential equations: a dynamical systems approach*. Springer-Verlag.

Chapter 4: Computer solution methods

A succinct introduction to these methods, for both ordinary and partial differential equations, can be found in Chapter 20 of Kreyszig [11].

For a more extensive treatment:

17. Faires, J. D. and Burden, R. L. (1993). *Numerical methods*. PWS-Kent.
18. Gerald, C. F. and Wheatley, P. O. (1994). *Applied numerical analysis*, (5th edn). Addison Wesley.

At a more advanced level still, but with an interesting historical perspective, try

19. Hairer, E., Nørsett, S. P. and Wanner, G. (1993). *Solving ordinary differential equations I*. Springer-Verlag.

A proof of the validity of the Runge–Kutta method may be found in [15]. It proceeds in the same spirit as Ex. 4.6, but at a much more advanced level, and substantial parts of the algebra are left to the reader.

While we have emphasized the flexibility and satisfaction of a 'do-it-yourself' approach using QBasic (see Appendices), it is of course possible to numerically integrate systems of ordinary differential equations using professional software instead. For PCs, try:

20. Koçak, H. (1989). *Differential and difference equations through computer experiments*, (2nd edn). Springer-Verlag.
21. Korsch, H. J. and Jodl, H. J. (1994). *Chaos: a program collection for the PC*. Springer-Verlag.

and for Macintoshes:

22. Hubbard, J. H. and West, B. H. (1993). *MacMath: a dynamical systems software package for the Macintosh*. Springer-Verlag.

The computer algebra packages MATHEMATICA and MAPLE contain routines for the numerical solution of differential equations, and in this connection try

23. Coombes, K. R., Hunt, B. R., Lipsman, R. L., Osborn, J. E. and Stuck, G. J. (1995). *Differential equations with Mathematica*. Wiley.
24. Kreyszig, E. and Normington, E. J. (1994). *Maple computer manual for Advanced Engineering Mathematics (7th edn)*. Wiley.

Chapter 5: Elementary oscillations

For a general introduction to mechanics try

25. Smith, P. and Smith, R. C. (1990). *Mechanics*, (2nd edn). Wiley.
26. Marion, J. B. and Thornton, S. (1995). *Classical dynamics of particles and systems*, (4th edn). Saunders College Publishing.
27. Lunn, M. (1991). *A first course in mechanics*. Oxford University Press.

For more specific treatments of oscillations see also

28. French, A. P. (1971). *Vibrations and waves*. Chapman and Hall.
29. Pippard, A. B. (1989). *The physics of vibration*. Cambridge University Press.

A derivation of the formulae quoted in respect of the system in Fig. 5.13 can be found in an article by M. J. Moloney in the *American Journal of Physics*, Vol. 46, pp. 1245–1246 (1978).

Chapter 6: Planetary motion

For semi-'popular' historical accounts:

30. Cohen, I. B. (1985). *The birth of a new physics*. Penguin.
31. Peterson, I. (1993). *Newton's clock; chaos in the Solar System*. Freeman.

The textbooks [25]–[27] contain good introductions to the theory. At a slightly more advanced level try:

32. Landau, L. D. and Lifshitz, E. M. (1976). *Mechanics*, (3rd edn). Pergamon.
33. Meirovitch, L. (1970). *Methods of analytical dynamics*. McGraw-Hill.

The second of these contains an introduction to the three-body problem, but in this connection see also

34. Roy, A. E. (1978). *Orbital motion*. Adam Hilger.
35. Szebehely, V. (1967). *Theory of orbits*. Academic Press.
36. Boccaletti, D. and Pucacco, G. (1996). *Theory of orbits*. Springer-Verlag.

Finally, the following book/software package contains some computer simulations of orbital motion:

37. Hawkins, B. and Jones, R. S. (1995). *Classical mechanics simulations*. Wiley.

Chapter 7: Waves and Diffusion

Further reading on partial differential equations requires some knowledge of the calculus of functions of several variables, as in, say, [8] or [9].

A good introduction to partial differential equations may be found in the relevant chapter of Kreyszig [11]. For a more extensive treatment try

38. Strauss, W. A. (1992). *Partial differential equations*. Wiley.

For appropriate numerical methods see Chapter 20 of Kreyszig [11], the relevant chapters of [17] and [18], and

39. Morton, K. W. and Mayers, D. F. (1994). *Numerical solution of partial differential equations*. Cambridge University Press.

For more on differential equations applied to biology see

40. Murray, J. D. (1989). *Mathematical biology*. Springer-Verlag.
41. Jones, D. S. and Sleeman, B. D. (1983). *Differential equations and mathematical biology*. George Allen & Unwin.

Chapter 8: The best of all possible worlds?

Two semi-'popular' accounts of minimum principles in Nature are

42. Hildebrandt, S. and Tromba, A. (1996). *The parsimonious universe: shape and form in the natural world*. Copernicus.
43. Isenberg, C. (1992). *The science of soap films and soap bubbles*. Dover.

For the calculus of variations and its applications to mechanics see the relevant chapter of [12] or [26], or Chapter 2 of

44. Hildebrand, F. B. (1992). *Methods of applied mathematics*. Dover.

For further reading on Lagrange's equations of motion try [26], [27], [32], [33] or

45. Woodhouse, N. M. J. (1987). *Introduction to analytical dynamics*. Oxford University Press.

Chapter 9: Fluid flow

For this subject I recommend, not surprisingly,

46. Acheson, D. J. (1990). *Elementary fluid dynamics*. Oxford University Press.

but try also

47. Tritton, D. J. (1988). *Physical fluid dynamics*. Oxford University Press.

For an inspiring collection of photographs of fluid motion see

48. van Dyke, M. (1982). *An album of fluid motion*. Parabolic Press.

A largely non-mathematical account of fluid flow in relation to aeroplane flight can be found in

49. Sutton, Sir Graham (1965). *Mastery of the air*. Hodder and Stoughton.

Chapter 10: Instability and Catastrophe

At a 'popular' level:

50. Stewart, I. and Golubitsky, M. (1992). *Fearful symmetry*. Penguin.

At a more advanced level:

51. Pippard, A. B. (1985). *Response and stability: an introduction to the physical theory*. Cambridge University Press.
52. Thompson, J. M. T. (1982). *Instabilities and catastrophes in science and engineering*. Wiley.
53. Saunders, P. T. (1980). *An introduction to catastrophe theory*. Cambridge University Press.

Chapter 11: Non-linear oscillations and chaos

At a 'popular' level:

54. Gleick, J. (1993). *Chaos*. Abacus.
55. Stewart, I. (1990). *Does God play dice*? Penguin.
56. Hall, N. (ed.) (1991). *The New Scientist guide to chaos*. Penguin.

Slightly more advanced:

57. Lorenz, E. N. (1993). *The essence of chaos*. UCL Press.
58. Ruelle, D. (1993). *Chance and chaos*. Penguin.

Some good general texts on non-linear systems:

59. Strogatz, S. H. (1994). *Non-linear dynamics and chaos*. Addison Wesley.
60. Jordan, D. W. and Smith, P. (1987). *Non-linear ordinary differential equations*, (2nd edn). Oxford University Press.
61. Drazin, P. G. (1992). *Nonlinear systems*. Cambridge University Press.
62. Grimshaw, R. (1990). *Nonlinear ordinary differential equations*. Blackwell Scientific.
63. Bender, C. M. and Orszag, S. A. (1978). *Advanced mathematical methods for scientists and engineers*. McGraw-Hill.

Some texts specifically on chaotic dynamics:

64. Baker, G. L. and Gollub, J. P. (1996). *Chaotic dynamics: an introduction*, (2nd edn). Cambridge University Press.
65. Moon, F. C. (1992). *Chaotic and fractal dynamics: an introduction for applied scientists and engineers*. Wiley-Interscience.
66. Mullin, T. (ed.) (1993). *The nature of chaos*. Oxford University Press.
67. Peitgen, H.-O., Jürgens, H. and Saupe, D. (1992). *Chaos and fractals, new frontiers of science*. Springer-Verlag.
68. Thompson, J. M. T. and Stewart, H. B. (1986). *Nonlinear dynamics and chaos*. Wiley.

Chapter 12: The not-so-simple pendulum

Stephenson's original paper on the inverted pendulum appeared in the *Memoirs and Proceedings of the Manchester Literary and Philosophical Society*, Vol. 52 (8), pp. 1–10 (1908).

Chaotic motions of single pendulums driven in various ways are examined in [64]–[66]. Baker and Gollub, in particular, build much of their exposition around a chaotic pendulum. Korsch and Jodl [21] have an interesting chapter on a chaotic double pendulum which is *un*driven and also undamped.

Chapter 12 includes some of my own research. The inverted pendulum theorem of Section 12.5 first appeared in *Proceedings of the Royal Society* A, Vol. 443, pp. 239–245 (1993), and the upside-down dancing oscillations of Figs 12.2, 12.3 were reported later in the same journal, Vol. 448, pp. 89–95 (1995). The experiments on upside-down multiple pendulums were reported by Tom Mullin and myself in *Nature*, Vol. 366, pp. 215–216 (1993).

We mentioned in that paper one puzzling lack of agreement between the theorem and the experiment which has since been resolved. The theoretical upper limits to ε/l quoted in the third sentence from the end of the paper are incorrect, and should be replaced by 0.18 for the double pendulum and 0.095 for the triple pendulum. This error arose because we omitted to take proper account of the substantial mass of the bearings which joined the rods, and I would like to take this opportunity of thanking Dr. P. Jaeckel for identifying the source of the difficulty.

The apparatus itself was constructed by Keith Long, who can (just) be seen in Figs. 12.15(a) and 12.18(b).

At the time the experiments were done, Tom Mullin was at the Clarendon Laboratory, Oxford, but he is currently in the Department of Physics and Astronomy at the University of Manchester.

Appendix A: Elementary programming in QBasic

A.1 Introduction

These notes are intended to help the reader get the various computer programs in this book up and running with the minimum of practical difficulty. *No previous computing experience is required.*

The programming language QBasic, developed by the Microsoft Corporation, has been chosen in this book because:

(i) it is well suited to our needs, yet relatively easy to learn, particularly for beginners;

(ii) it has its own built-in graphics facilities, so that we may display results in graphical, or even animated, form, without the need for additional software;

(iii) it has, for some years now, been distributed with Microsoft's operating systems (i.e., MS-DOS or WINDOWS 95), which are used in a great many personal computers.

All the reader should need, therefore, is access to an IBM-compatible PC running on MS-DOS *or* WINDOWS 95. Many of these now have an Intel 486DX or Pentium processor, but a 386-based computer running on MS-DOS will be fast enough for most of our purposes, particularly if it has a maths coprocessor.*

I hope that the notes which follow will be more generally helpful, but they are aimed mainly at complete beginners. I bid a particularly warm welcome to any reader who actually *dislikes* the prospect of having to do some mathematical computing; this was the author himself, just a few years ago.

* Program execution on a given computer can generally be speeded up by purchasing a *compiled* version of QBasic, namely Microsoft Quick Basic 4.5 or Microsoft Visual Basic for DOS; in my experience the programs in this book then run some 5 or 6 times faster, for a given time step h.

A.2 Getting started

The first step is to call up QBasic—assuming, of course, that it is on your computer. Be prepared to seek help in doing this, if necessary, but in the first instance simply switch on the machine.

If you are greeted by the MS-DOS prompt

```
C:\>
```

you can typically get into QBasic by typing `qbasic` and then pressing `RETURN` (or `ENTER`).

You may, on the other hand, find that your computer runs some version of WINDOWS when you switch it on. For the notes which follow, it is then probably simplest to *exit* WINDOWS and run the computer in MS-DOS mode instead, seeking help, again, if necessary.

If your computer has WINDOWS 95 as its fundamental operating system it is possible that QBasic may not be immediately in evidence; the manufacturer may not have pre-loaded the whole of WINDOWS 95. However, QBasic *is* included in the *complete* version of WINDOWS 95, and on the CD-ROM, for example, QBasic can be located by browsing the CD, double-clicking with the mouse on 'other', and then double-clicking again on 'oldmsdos'.

One way or another, you should eventually receive the message 'Welcome to MS-DOS QBasic', and on pressing the `ESC` key as advised you should obtain the kind of screen shown in Fig. A.1.

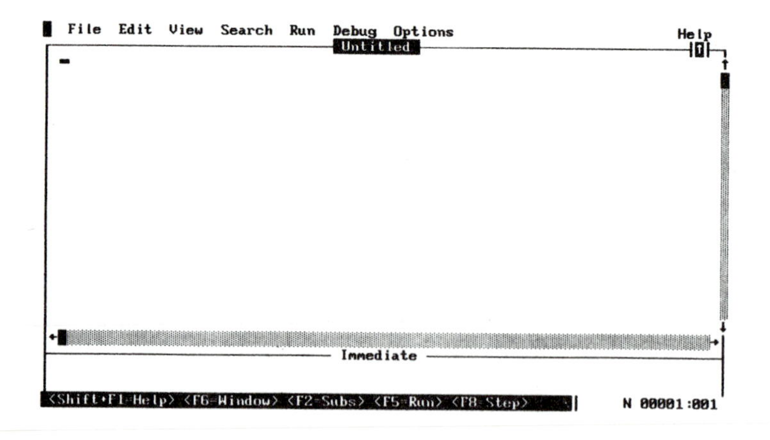

Fig. A.1 The QBASIC screen.

You are now ready to type in and run a program, and from this point on, with luck, the following notes should be all you need.

As a simple start, type in

```
CLS
a = 3
b = 4
x = a + b
PRINT x
```

pressing the RETURN (or ENTER) key after each line. Here CLS means 'clear the screen' and PRINT means 'display on screen'.

You may correct mistakes in typing the program—or otherwise edit it—by using the ← ↑ ↓ → keys to move the cursor to the relevant position on the screen. Press the DELETE key to delete a character at the current cursor location.

To *run* the program press the key sequence Alt-R followed by S, where *by the notation* Alt-R *we mean pressing the* R *key while holding down the* Alt *key.*

You should now find that the number 7 appears on the screen, as one would hope. Press any key to return to the program. We will now edit it to introduce one or two useful additional features of QBasic.

First, we may use a colon to compress the second and third lines into the single line

```
a = 3 : b = 4
```

The INPUT command allows us to assign numerical values to variables without changing the program itself. Thus if we replace the above line with

```
INPUT "Two numbers"; a,b
```

and run the program we receive the prompt

```
Two numbers?
```

If we type these in, separated by a comma, and press RETURN, their sum x appears on screen.

The LOCATE command allows us to control the position of both input and output on screen, and works by dividing the screen into a number of rows (either 1–25 or 1–30, depending on the screen resolution) and columns (1–80), starting from the top left-hand corner. Thus

```
CLS
LOCATE 6,6 : INPUT "Two numbers"; a,b
x = a + b
LOCATE 20,50 : PRINT "Their sum is"; x
```

causes the input and output to appear at very different parts of the screen.

Now *save* the program currently on screen by inserting a disk into the floppy disk drive and pressing `Alt-F` followed by `A` (for Save *A*s). Then type in `a:sum` and press `RETURN`. This will save the program on the disk under the title `SUM.BAS`. Alternatively, the program could have been saved on the hard disk inside the computer itself by typing `c:sum` instead.

Now leave QBasic altogether by typing `Alt-F` followed by `X` (for E*x*it). This will return you to your original starting point, whether that be the MS-DOS prompt or otherwise.

Let us now resume programming by calling up QBasic as before.

Type `Alt-F` followed by `O` (for *O*pen), and in response to the request for a FileName type `a:` and press `RETURN`. The directory of your floppy disk will appear, and by typing `SUM` and pressing `RETURN` you can, if you wish, call up that program again.

Instead, press `ESC` to clear, and then type in and run the new program

```
CLS
t = 1
t = t + 0.1
PRINT t
```

There is nothing 'wrong' with the third line here, because the equals sign = has a different meaning from the usual mathematical one: *the program line*

x = right-hand side

means 'calculate the current numerical value of the right-hand side and set `x` *equal to that value'*. For this reason only a single variable should appear on the left-hand side; QBasic will not know what to make of

```
a + b = x,
```

and will complain accordingly.

Being able to *stop* is almost as important as getting started, so we round off this section with a few notes on the matter.

To stop a QBasic program *while it is running*, because it will otherwise go on too long (perhaps for ever), press `CTRL-Break`, unless you have already made provision within the program itself for interrupting it in some other way.

Pressing `Pause`, on the other hand, will cause the program to stop *temporarily*, permitting inspection of what is on screen at the time. Pressing any key will then cause the program to continue running again.

The `ESC` key can be quite helpful, in various circumstances, in getting something off the screen so that you get 'back to where you were'. Guidance on how to do this may, however, be somewhere up on screen, near one of the borders. We have already noted, for instance, that if a QBasic program is up on screen we may leave QBasic altogether by pressing `Alt-F` and then `X`.

In sufficiently dire circumstances, perhaps as a result of pressing wrong keys by mistake, reboot the computer by holding down *both* `CTRL` *and* `ALT` and then pressing `DELETE`.

In the event of a complete 'crash', when no input from the keyboard leads to any response on screen at all, take out any disk that may be in the `a:` drive, and then turn the computer off (and on again).

A.3 Mathematical variables, operations, and functions

Variables, such as `x`, or `a6`, or `SPEED`, and numerical constants, such as 6.9143, are typically stored in the computer's memory with *single precision*, i.e. as a seven-digit number (of which only six may be accurate, as a result of rounding errors).

The *double precision* instruction `DEFDBL`, together with the symbol `#` after each numeric constant which occurs in the program as part of the actual calculation, causes the computer to work with much higher accuracy:

```
DEFDBL a, x
a = 0.3#
x = a + 0.4#
PRINT x
```

More generally, `DEFDBL q` means treat any variable whose name *begins* with `q` as a double precision variable. This `DEFDBL` instruction by itself is not, however, enough; rerun the above program with one or both of the `#` characters missing, and note the difference in the result.

An *integer* variable, say `x`, can be always accurately stored as an integer, rather than as a single precision variable, by defining it as such early in the program:

```
DEFINT x
```

An integer numerical constant occurring in the program will be stored accurately as such if the character `%` is put immediately after it.

The standard mathematical operations are represented in QBasic as follows:

$a+b$	`a + b`
$a-b$	`a - b`
$a\times b$	`a * b`
a/b	`a/b`
a^b	`a ^ b`

Within a given expression operations are performed in the order: exponentiations (a^b), multiplication and division, addition and subtraction—unless that order is changed by the use of brackets. Thus

```
c = a ^ 2 + b/c + d
```

will be understood as

$$c = a^2 + \frac{b}{c} + d,$$

and if our intention was

$$c = \frac{a^2 + b}{c + d}$$

we should have used brackets as follows:

```
c = (a ^ 2 + b)/(c + d).
```

Various standard functions are available in QBasic, including

```
SQR(x)
```

which gives the positive square root of the positive number x. (QBasic cannot cope directly with complex numbers.) Other functions include

```
SIN(x),    COS(x),    TAN(x),
```

x being *in radians* if it denotes some angle. The functions

```
EXP(x),    LOG(x)
```

denote e^x and $\log x$, respectively.

The argument of a function need not be a single number or a single variable. Thus

```
X = (-B + SQR(B ^ 2 - 4 * A * C))/(2 * A)
```

will give one of the two roots of $ax^2 + bx + c = 0$, provided of course that numerical values for a, b, and c have been specified earlier in the program and that $b^2 > 4ac$.

Other 'standard' functions which are useful include:

`ABS(X)`	The absolute value of x, $\|x\|$.
`SGN(X)`	1 if $x > 0$, -1 if $x < 0$, and 0 if $x = 0$.
`INT(X)`	The integer part of x, i.e. the greatest integer which is less than or equal to x.

The function `INT(X)`, for instance, can be used for rounding off a number to, say, two decimal places, for convenience of display on screen:

```
X = INT(100 * X + .5)/100
```

The function `CSNG(x)` converts a double precision number `x` to its single precision equivalent.

User-defined functions can be convenient. Thus

```
DEF FNF(x) = 1/(x ^ 2 + 1)
```

defines the function $f(x) = 1/(x^2 + 1)$, and `FNF(X)` may be employed, later in the program, in the same way as any of the standard functions such as `SIN(X)`. A user-defined function may have more than one variable;

```
DEF FNR(X,Y) = SQR(X ^ 2 + Y ^ 2)
```

defines $r(x, y) = \sqrt{(x^2 + y^2)}$.

A.4 Program loops

Simple *loops* lie at the heart of the main programs which follow in Appendix B.

(a) `DO...LOOP`

Here is an example:

```
CLS
t = 0 : h = 0.05
 DO
  x = t ^ 2
  PRINT t,x
  t = t + h
 LOOP UNTIL t > 1
```

The indentation of various lines here is ignored by the computer, but makes the program easier to read, by highlighting its structure. The program provides $x = t^2$ for values of t between 0 and 1 at intervals of $h = 0.05$, because the `DO...LOOP` causes the contents of that loop to be executed again and again until the `UNTIL` condition is met. (When the program is actually run, the effect of single precision rounding errors can be quite apparent.)

If the last program line were simply `LOOP` instead, the program would go on running forever. It could, however, be stopped at any time by pressing `CTRL-Break`, as we pointed out at the end of Section A.2.

A 'softer' way of stopping the program running when we feel like it is to replace the last line by, say,

```
LOOP UNTIL INKEY$ = "q".
```

Pressing the `q` key, which we have chosen for its connotations of 'quit', will then stop the program running.

`DO...LOOPS` may be *nested*, one inside another, and the inner loop is then executed repeatedly, until its end-condition is met, during *each* passage through the outer loop. We make frequent use of such nested loops in Appendix B.

(b) `FOR...NEXT`

This type of loop ends not when some condition is at last met, but when the loop has been executed a predetermined number of times. An example is

```
CLS
s = 0
 FOR N = 1 TO 100
  s = s + 1/N ^ 2
  PRINT N,s
 NEXT
```

which generates the sum of the first 100 terms of the famous series

$$\sum_{n=1}^{\infty} \frac{1}{n^2} = \frac{\pi^2}{6}.$$

(c) `IF...THEN`

The `IF...THEN...END IF` structure can provide additional flexibility within a program, particularly in combination with a loop. The following program is a simple example:

```
CLS
ymin = 100
 FOR x = 1 TO 10 STEP 0.001
  y = x + 2/x
   IF y<ymin THEN
    ymin = y : xvalue = x
   END IF
 NEXT
PRINT xvalue, ymin
```

and tries (inefficiently) to find the least value of $y = x + 2/x$ in the interval $1 < x < 10$, and the value of x, `xvalue`, which gives it.

A.5 Graphics

Our first step in using QBasic graphics is to clear the screen and specify high resolution:

```
CLS : SCREEN 9
```

The computer then regards the screen as being divided into 640×350 *pixels*, and labels them according to the screen coordinate system shown in Fig. A.2.* To specify which part of the screen we wish to use for graphics we then type in, for example,

```
VIEW (100,100) - (500,200),0,4.
```

* In working throughout with `SCREEN 9` we are assuming that the reader's computer is set up to run high-resolution graphics if required. In fact, many modern PCs will support higher screen resolution still; `SCREEN 12` divides the screen into 640×480 pixels, and various numbers which we give in `VIEW` and `LOCATE` commands then need to be changed accordingly (see Appendix B).

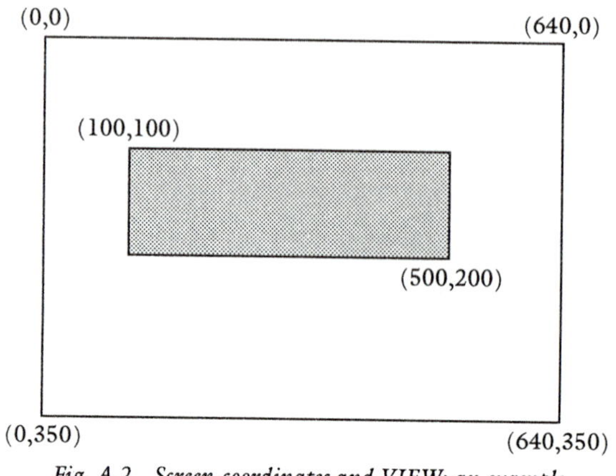

Fig. A.2 Screen coordinates and VIEW: an example.

Here the first pair of '*screen*' coordinates denotes the top left corner of the viewing area and the second pair denotes the bottom right corner. The 0 which follows calls for a black background in the viewing area, while the 4 calls for a red border round it. Other numbers give different colours, according to the following scheme:

0	Black	8	Grey
1	Blue	9	Light blue
2	Green	10	Light green
3	Cyan	11	Light cyan
4	Red	12	Light red
5	Magenta	13	Light magenta
6	Brown	14	Yellow
7	White	15	High-intensity white

The next step is to place our own *mathematical* coordinate system as we want it in the viewing area. To arrange this as in Fig. A.3, for example, we type in

```
WINDOW (0, -5) - (30,5)
```

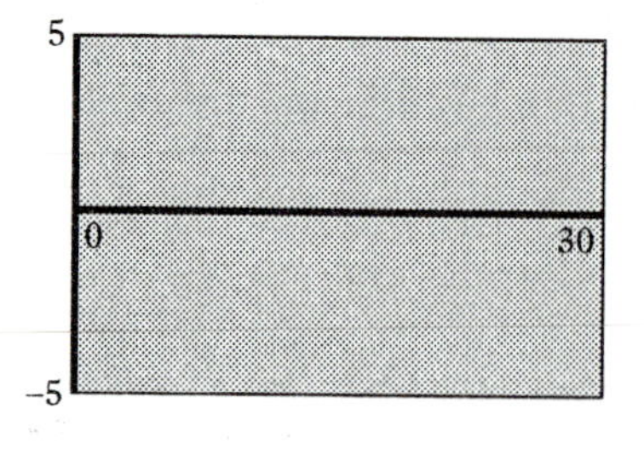

Fig. A.3 Mathematical coordinates and WINDOW.

the first pair being the desired (mathematical) coordinates of the lower left corner and the second being the corresponding mathematical coordinates of the upper right corner. We may then draw in the axes with

```
LINE (0,-5) - (0,5), 9
LINE (0,0) - (30,0), 9
```

The 9 will cause the axes to appear in light blue.

The final step is to actually plot a point on screen, which is done with the command PSET, followed by the two mathematical coordinates in question, in brackets, and separated by a comma. Thus if we follow the above five lines of program with

```
h = 0.01
t = 0
DO
 x = 3 * SIN(t)
 PSET (t, x), 14
 t = t + h
LOOP UNTIL t> 30
```

we obtain a sine curve on screen, in yellow.

Two other simple graphics commands, CIRCLE and PAINT, are occasionally useful. As an illustration, take just the CLS, VIEW, and WINDOW lines above and add

```
CIRCLE (15,0), 3,14
```

This will give a circle, centered on the point with mathematical coordinates (15, 0), radius 3, in yellow (14). The further line

```
PAINT (15,0), 14
```

is an instruction to paint outwards, from the point (15, 0), in yellow, until a closed, yellow boundary is encountered. In this way the interior of the circle of radius 3 will be painted yellow.

Appendix B: Ten programs for exploring dynamics

Introduction

We now consider 10 QBasic programs which are simple to understand yet powerful enough to allow us to explore some interesting dynamical questions. One advantage of this approach over the use of highly sophisticated software is that it is possible to keep good track of what the computer is actually doing, so that *we* can justifiably take satisfaction from solving the problem.

The programs may be typed in line-by-line or downloaded freely from my web site **http://www.jesus.ox.ac.uk/~dacheson.**

The first six are general-purpose programs for systems of ordinary differential equations, the next three are variations on these for particular dynamical problems, and the last relates to a *partial* differential equation, namely the diffusion equation (7.12). The notes which accompany them assume that the reader is familiar with Appendix A and the short differential equation programs in Chapter 4.

Graphics window shape

Some of the programs, as written, require a *square*-shaped viewing area for graphics on screen if we are to avoid a 'squashed' representation of the 'actual' motion. This is true, for instance, of the programs `PENDANIM`, `PENDOUBL`, and `THREEBP`, as well as `NPHASE` when that is used, for example, to compute planetary orbits (Fig. 6.11). The command `VIEW (180,17)-(595,330)` is meant to achieve a square viewing area, and does so on my own PC, but the reader may need to change these numbers a little to get the same effect on his or her own machine. In using `VIEW (a,b)-(c,d)` the key quantity controlling the *shape* of the rectangular graphics window is, of course, the ratio $(d-b)/(c-a)$.

Screen resolution

All programs use `SCREEN 9` high-resolution graphics, but may be modified easily to `SCREEN 12` very-high-resolution graphics if desired. Statements such as `LOCATE (13,1)` should then be changed by adding 2 or 3 to the first coordinate in brackets. The `VIEW` instructions will also need to be changed; in place of `(180,17)-(595,330)` in `screen 9` one might try `(160,20)-(590,450)` in `screen 12`, while `(20,30)-(575,240)` could be replaced similarly by `(20,35)-(575,323)`. The numbers here are, however, only guidelines, for the reasons given in the paragraph above.

Minimizing 'overflow'

Most of the programs contain nothing to limit the magnitudes of the variables involved, so some runs will end with an 'Overflow' message. Readers who wish to minimize this, and indeed the possibility of attendant 'crashes', may wish to insert conditions on variable size in the relevant `DO...LOOP`s, and the program `1XT` provides one example of how to do this.

In the 10 programs which follow, the parts which need to be changed most frequently during normal use have been highlighted.

Program 1: `1XT`

```
REM ****** Setting up graphics ******

  CLS : SCREEN 9: PAINT (1, 1), 9
  xm = 3: tm = 5
  VIEW (180, 17) - (595, 330), 0, 13
  WINDOW (0, -xm) - (tm, xm)
  LINE (0, 0) - (tm, 0), 11: LINE (0, -xm) - (0, xm), 11
  LOCATE 12, 76: PRINT tm: LOCATE 1, 21: PRINT xm

REM ****** Function f(x, t) ******

  DEF fnf (x, t) = (1 + t) * x + 1 - 3 * t + t ^ 2

REM ****** Direction field ******

p = 25
  FOR x = xm TO -xm STEP -2 * xm / p
    FOR t = 0 TO tm STEP tm / p
      x1 = fnf(x, t) / xm: t1 = 2 / tm
      s = p * SQR(x1  ^  2 + t1  ^  2)
      x2 = fnf(x, t) / s: t2 = 1 / s
      LINE (t, x) - (t + t2, x + x2), 9
      CIRCLE (t + t2, x + x2), .003 * tm, 9
    NEXT t
  NEXT x
```

```
REM ****** Step-by-step method ******

DO
  t = 0: LOCATE 13, 1: INPUT "x0"; x
  h = .01
    DO
       GOSUB Runge
       t = t + h
       PSET (t, x), 14
    LOOP UNTIL ABS(t - tm) < h / 2 OR ABS(x) > xm
      LOCATE 19, 1: PRINT "t="; t
      LOCATE 20, 1: PRINT "x="; x
LOOP

REM ****** Subroutines ******

Euler:

    c1 = h * fnf(x, t)
    x = x + c1
    RETURN

ImpEuler:

    c1 = h * fnf(x, t)
    c2 = h * fnf(x + c1, t + h)
    x = x + (c1 + c2) / 2
    RETURN

Runge:

    c1 = h * fnf(x, t)
    c2 = h * fnf(x + c1 / 2, t + h / 2)
    c3 = h * fnf(x + c2 / 2, t + h / 2)
    c4 = h * fnf(x + c3, t + h)
    x = x + (c1 + 2 * c2 + 2 * c3 + c4) / 6
    RETURN
```

This program is for any initial value problem of the type (4.2), i.e.

$$\frac{dx}{dt} = f(x, t), \qquad x = x_0 \text{ when } t = 0.$$

On running, the program:

(i) draws the *direction field*;
(ii) asks for an initial value x_0, and then plots the solution curve *by any one of the three standard step-by-step methods*;
(iii) asks for another x_0, and plots the new solution curve, and continues to do this so that we may see *several such curves together on screen*.

The remark (`REM`) lines, like the various blank lines and indentations, are ignored by the computer when the program is run, and are included simply to make the program more easy to follow.

The `PAINT` command in the first line is purely cosmetic, but gives a pleasing blue (`9`) background on screen. In setting up the graphics we specify a portion of the screen (`VIEW`), set up our own t, x coordinates in that portion (`WINDOW`) and then draw and label the axes.

The function $f(x, t)$ is defined next; the particular one in the program as written is that occurring in eqn. (4.1).

We then plot the direction field. The number of points at which the 'flow' direction is indicated is determined by the parameter `p`, while `s` controls the length of each short line segment occurring on screen; both may be amended, if desired. The `CIRCLE` command, with a radius which is too small for the screen to resolve properly, is a lazy substitute for putting a proper arrow at one end of each line segment.

The heart of the program, as ever, is the step-by-step procedure, and here one `DO...LOOP` is inside another. The inner `DO...LOOP UNTIL` differs from those in the programs of Chapter 4 by hiving off the updating process to a **subroutine** by the command `GOSUB`. As written, the computer switches to the subroutine called `Runge` at this point, and as soon as it encounters `RETURN` it goes back to the main program, at the line (`t=t+h`) immediately after the `GOSUB`. To use, say, the `Euler` subroutine here instead, simply replace the line `GOSUB Runge` by `GOSUB Euler`.

The inner `DO...LOOP UNTIL` ends when *either* `t` reaches the specified end-time `tm` *or* $|x|$ becomes too large for the point (t, x) to appear in the graphics window. If we do not include `OR ABS(x) > xm` and `INPUT`, say, `x0=-1`, then the run is terminated by an 'overflow' message instead.

The whole inner loop is within an outer `DO...LOOP` which has the effect of resetting `t=0` when the solution curve has been plotted and then asking (`INPUT`) for a new initial value x_0.

Program 2: 2PHASE

```
REM ****** Setting up graphics ******

CLS : SCREEN 9: PAINT (1, 1), 9
xm = 4: ym = 4: tm = 25
VIEW (180, 17) - (595, 330), 0, 13
WINDOW (-xm, -ym) - (xm, ym)
LINE (-xm, 0) - (xm, 0), 11: LINE (0, -ym) - (0, ym), 11
LOCATE 13, 76: PRINT xm: LOCATE 1, 48: PRINT ym

REM ****** Functions f(x,y) and g(x,y) ******

  DEF fnf (x, y) = y
  DEF fng (x, y) = -SIN(x)

REM ****** Direction field ******

p = 25
    FOR y = ym TO -ym STEP -2 * ym / p
      FOR x = -xm TO xm STEP 2 * xm / p
        x1 = fnf(x, y) / xm: y1 = fng(x, y) / ym
        s = p * SQR(x1 ^ 2 + y1 ^ 2)
        x2 = fnf(x, y) / s: y2 = fng(x, y) / s
        LINE (x, y) - (x + x2, y + y2), 9
        CIRCLE (x + x2, y + y2), .003 * xm, 9
      NEXT x
    NEXT y

REM ****** Step-by-step method ******

  DO
     t = 0: LOCATE 13, 1: INPUT "x0,y0"; x, y
     h = .01
    DO
        GOSUB ImpEuler
        t = t + h
        PSET (x, y), 14
    LOOP UNTIL INKEY$ = "q"
  LOOP

REM ****** Subroutines ******

Euler:
```

```
    c1 = h * fnf(x, y)
    d1 = h * fng(x, y)
    x = x + c1
    y = y + d1
    RETURN

ImpEuler:

    c1 = h * fnf(x, y)
     d1 = h * fng(x, y)
    c2 = h * fnf(x + c1, y + d1)
     d2 = h * fng(x + c1, y + d1)
    x = x + (c1 + c2) / 2
     y = y + (d1 + d2) / 2
    RETURN
```

This has much the same structure and purpose as `1XT`, but is for the *second-order autonomous* system

$$\frac{dx}{dt} = f(x, y),$$

$$\frac{dy}{dt} = g(x, y),$$

with $x = x_0$, $y = y_0$ when $t = 0$.

The direction field, and individual solution paths, are plotted *in the x, y phase plane*. The use of `INKEY$` in the program at the end of the inner `DO...LOOP` means that the computer keeps plotting each solution path until the `q` key is pressed. Again, several such paths can be seen on screen together, to compare the effects of different initial conditions.

The particular example shown has $f(x, y) = y$ and $g(x, y) = -\sin x$, so is for the system $\dot{x} = y$, $\dot{y} = -\sin x$, i.e. the (non-dimensional) simple pendulum equation $\ddot{x} + \sin x = 0$. (We are here denoting by x the angle which we denoted by θ in (5.43) and (5.46).)

Program 3: `2XTPHASE`

```
REM ****** Setting up graphics ******

   CLS : SCREEN 9: PAINT (1, 1), 9
    xm = 2: ym = 2: tm = 50
```

```
  VIEW (20, 30) - (575, 240), 0, 9
  WINDOW (0, -xm) - (2 * tm, xm)
    LINE (.8 * tm, -xm) - (.8 * tm, xm), 9
    LINE (tm, -xm) - (tm, xm), 9
    PAINT (.9 * tm, 0), 9
   LINE (0, 0) - (.8 * tm, 0), 11
   LINE (0, -xm) - (0, xm), 11
   LINE (tm, 0) - (2 * tm, 0), 11
   LINE (1.5 * tm, -xm) - (1.5 * tm, xm), 11
  LOCATE 10, 32: PRINT tm: LOCATE 2, 1: PRINT xm
  LOCATE 10, 74: PRINT xm: LOCATE 2, 53: PRINT ym
   COLOR 10: LOCATE 19, 17: PRINT "xt"
   LOCATE 19, 53: PRINT "phase": COLOR 15

REM ****** Step-by-step method ******

  w = 1: k = .1
  DEF fnf (x, y, t) = y
  DEF fng (x, y, t) = -k * y - w ^ 2 * x

    t = 0: LOCATE 20, 1: INPUT "x0,y0"; x, y
    h = .01

  DO
    GOSUB ImpEuler
    t = t + h
    PSET (.8 * t, x), 13
     IF x > -xm THEN
      PSET ((1.5 + .5 * x / xm) * tm, y * xm / ym), 14
     END IF
  LOOP UNTIL ABS(t - tm) < h / 2
     LOCATE 22, 1: PRINT "t="; t
     LOCATE 22, 20: PRINT "x="; x
     LOCATE 23, 20: PRINT "y="; y
  END

REM ****** Subroutines ******

Euler:

    c1 = h * fnf(x, y, t)
    d1 = h * fng(x, y, t)
    x = x + c1
    y = y + d1
    RETURN
```

```
ImpEuler:

    c1 = h * fnf(x, y, t)
    d1 = h * fng(x, y, t)
    c2 = h * fnf(x + c1, y + d1, t + h)
    d2 = h * fng(x + c1, y + d1, t + h)
    x = x + (c1 + c2) / 2
    y = y + (d1 + d2) / 2
    RETURN
```

This is for a second-order system which may or may not be autonomous:

$$\frac{dx}{dt} = f(x, y, t),$$

$$\frac{dy}{dt} = g(x, y, t),$$

with $x = x_0$, $y = y_0$ at $t = 0$.

The program plots *both x, y paths and t, x solution curves simultaneously*. Only one `WINDOW` command is used, the x, y and t, x plots being separated by two vertical blue lines, with the space between them painted blue. This explains the rather curious `PSET` coordinates for plotting a point in the x, y plane.

If the system is non-autonomous, with t occurring explicitly in either $f(x, y, t)$ or $g(x, y, t)$, then the x, y plane shows only a *projection* of the actual phase path, of course, for the phase space itself is then three-dimensional (see Section 3.6).

As written, the program is in fact for the autonomous system $\dot{x} = y$, $\dot{y} = -ky - \omega^2 x$, with parameters $\omega = 1$, $k = 0.1$. This corresponds to the damped linear oscillator equation (5.14), i.e. $\ddot{x} + k\dot{x} + \omega^2 x = 0$.

Program 4: `NPHASE`

```
DEFDBL A-H, K-M, O-Z: DEFINT I-J, N
n = 3: OPTION BASE 1
DIM x(n), xc(n), f(n), c1(n), c2(n), c3(n), c4(n)

REM ****** Setting up graphics ******

   CLS : SCREEN 9: PAINT (1, 1), 9
    xm = 30: ym = 60: tm = 15#
   VIEW (180, 17) - (595, 330), 0, 14
```

```
  WINDOW (-xm, -ym) - (xm, ym)
  LINE (-xm, 0) - (xm, 0), 11
  LINE (0, -ym) - (0, ym), 11
   LOCATE 13, 76: PRINT xm: LOCATE 1, 48: PRINT ym
   LOCATE 15, 1: PRINT "xi"
   LOCATE 23, 2: PRINT "Time"

REM ****** Step-by-step method ******

  LOCATE 13, 2: INPUT "r"; r

  t = 0#: xc(1) = 5#: xc(2) = 5#: xc(3) = 5#
  h = .003#

    DO
      GOSUB Runge
      t = t + h
      PSET (xc(1), xc(3)), 14
      LOCATE 23, 6: PRINT CSNG(t)
    LOOP UNTIL ABS(t - tm) < h / 2
      LOCATE 16, 1
       FOR i = 1 TO n: PRINT CSNG(xc(i)): NEXT
    END

REM ****** Subroutines ******

Equations:

 f(1) = 10# * (x(2) - x(1))
 f(2) = -x(1) * x(3) + r * x(1) - x(2)
 f(3) = x(1) * x(2) - 8# * x(3) / 3#
 RETURN

Runge:

   FOR i = 1 TO n: x(i) = xc(i): NEXT
    GOSUB Equations
   FOR i = 1 TO n: c1(i) = h * f(i): NEXT

   FOR i = 1 TO n: x(i) = xc(i) + c1(i) / 2#: NEXT
    GOSUB Equations
   FOR i = 1 TO n: c2(i) = h * f(i): NEXT

   FOR i = 1 TO n: x(i) = xc(i) + c2(i) / 2#: NEXT
    GOSUB Equations
   FOR i = 1 TO n: c3(i) = h * f(i): NEXT
```

```
    FOR i = 1 TO n: x(i) = xc(i) + c3(i): NEXT
     GOSUB Equations
    FOR i = 1 TO n: c4(i) = h * f(i): NEXT

 FOR i = 1 TO n
  xc(i) = xc(i) + (c1(i) + 2# * c2(i)
          + 2# * c3(i) + c4(i)) / 6#      (as one line)
 NEXT

 RETURN
```

This program is the backbone for many others in this book, and it is for the *nth order* system

$$\dot{x}_1 = f_1(x_1, x_2, \ldots, x_n)$$

$$\vdots$$

$$\dot{x}_n = f_n(x_1, x_2, \ldots, x_n).$$

Note that the system is *autonomous*, but this is really no restriction, because we can turn any $(n-1)$th-order *non*-autonomous system into an nth-order autonomous one by adding the extra 'dependent' variable $x_n = t$ and the extra equation $\dot{x}_n = 1$ (see Section 3.6).

The program integrates these equations by the Runge–Kutta method and plots a two-dimensional projection of the phase path, which is itself in n-dimensional space.

The order of the system, n, is specified at the beginning of the second line of the program, and the right-hand sides of the equations, $f_1, f_2, \ldots, f_n$ are entered through a subroutine called `Equations`, the last highlighted section of `NPHASE`. Here `x(i)` denotes x_i, and so on, and *arrays* such as `x(1)`, `x(2)`,..., `x(n)` are all dimensioned (`OPTION BASE 1, DIM`) in lines 2 and 3.

Because we use the symbols `x(1)`,... to enter the right-hand sides of the equations, we need a different set for the actual variables which are given initially and are then continually updated and displayed on screen; these are denoted by `xc(1)`, `xc(2)`,..., `xc(n)`, as can be seen from the (highlighted) initial conditions and the `PSET` command.

Each updating of `xc(i)`, for `i=1` to `n`, works, then, by calling on the subroutine `Runge`, which in turn calls *four times* on the subroutine `Equations` in order to help with the calculation of `c1(i)`, `c2(i)`, `c3(i)`, and `c4(i)`, for `i=1` to `n`. These are finally combined in the usual way (cf. (4.36), and recall that the present system is *autonomous*) to produce a new array `xc(i)`, for `i = 1` to `n`.

The whole computation is carried out with double precision (`DEFDBL` in line 1, and `#` after all numerical constants involved in the actual calculation), but the final values of `xc(i)` are converted to single precision (`CSNG`) before being printed on screen.

As always, the most straightforward way of checking the whole calculation is to repeat it with the step size h reduced by a factor of 2. Only if the final values of `xc(i)` are almost unchanged can we place any confidence in the 'results', over the time interval in question.

As written, the program is for the Lorenz equations (11.7), and if we `INPUT r=28`, say, we obtain the distinctive 'butterfly' of Fig. 11.10.

Example: planetary motion

We may use `NPHASE` to integrate the fourth-order system (6.38) subject to the initial conditions (6.37).

To do this, change the highlighted parts of `NPHASE` to

```
n = 4
xm = 5.5 : ym = 5.5 : tm = 40#
    INPUT "v"; v
    xc(1) = 1# : xc(2) = 0# : xc(3) = 0# : xc(4) = v
h = 0.02#
PSET (xc(1),xc(2)), 12
f(1) = x(3)
f(2) = x(4)
f(3) = -x(1)/(x(1) ^ 2# + x(2) ^ 2#) ^ 1.5#
f(4) = -x(2)/(x(1) ^ 2# + x(2) ^ 2#) ^ 1.5#
```

This should be sufficient to confirm the results in Fig. 6.11.

To keep track of the *total energy* (6.40), which should be constant, add to 'Setting up graphics'

```
LOCATE 20,1 : PRINT "energy"
```

and add after the `PSET` command

```
kin = .5# * (xc(3) ^ 2# + xc(4) ^ 2#)
pot = -(xc(1) ^ 2# + xc(2) ^ 2#) ^ - .5#
energy = kin + pot
LOCATE 21,1 : PRINT energy
```

If we `INPUT v=1.3` we find that energy is conserved very well, but with `v=0.6` it changes rather more, particularly when the 'planet' is near the attractor at the origin and therefore moving rather fast on account of (6.12).

This can be seen more clearly by changing `xm` and `ym` to `1`. With `v=0.3` the whole step-by-step process goes wildly wrong near the origin, unless we reduce the time step greatly to `h=0.002`, or so.

A better way of curing this problem is to use a *variable time step h*. An effective but *ad hoc* way of doing this for the particular problem at hand is to replace the line `h=0.02#` by

```
hscale = 0.02#
```

and add just before the `LOOP...` statement

```
h = hscale * (xc(1) ^ 2# + xc(2) ^ 2#)
```

The time step then decreases in proportion to r^2 as the origin is approached, and with `v=0.3` the expected highly elliptical orbit is obtained, with energy conserved to a high degree of accuracy. In fact, quite acceptable results can be obtained with `hscale` as large as `0.1#`.

Finally, we should perhaps remark on the total disregard for computational *efficiency* in the above amendments to `NPHASE`; the quantity `xc(1) ^ 2# + xc(2) ^ 2#` is calculated anew much more often than it need be, during each time step `h`.

Example: the forced cubic oscillator

As a second example we take the *non-autonomous* equation (11.1) and recast it as

$$\begin{aligned} \dot{x} &= y, \\ \dot{y} &= -ky - x^3 + A\cos\Omega t, \\ \dot{t} &= 1, \end{aligned}$$

which is autonomous. We may then confirm Fig. 11.2 by changing the highlighted parts of `NPHASE` to

```
n = 3
xm = 4 : ym = 8 : tm = 50#
LOCATE 13, 2 : INPUT "a"; a
k = .05# : w = 1#
 xc(1) = 3# : xc(2) = 4# : xc(3) = 0#
h = .01#
PSET (xc(1), xc(2)), 14
f(1) = x(2)
f(2) = -k * x(2) - x(1) ^ 3# + a * cos(w * x(3))
f(3) = 1#
```

so that `x(1)` $=x$, `x(2)` $=y$, `x(3)` $=t$.

The paths we see on screen and in Fig. 11.2 are of course only two-dimensional projections of the actual phase paths, the phase space in this problem being three-dimensional.

A variation: NPHASEXT

This is a simple variation on NPHASE in which a t, x curve can be seen evolving on screen at the same time as the phase path.

Add to the original NPHASE

```
LINE (-xm, -.85 * ym) - (xm, -.85 * ym), 13
```

after the LINE statements in 'Setting up graphics', and add after the PSET statement

```
tinset = (2 * t/tm - 1) * xm
xinset = (.15 * xc(1)/xm - .85) * ym
PSET (tinset, xinset), 10
```

Another variation: NSENSIT

This is another simple variation on NPHASE in which the phase paths resulting from four different initial conditions can be seen evolving simultaneously.

The four n-vectors being continually updated are called xcc(1, i), xcc(2, i), xcc(3, i), xcc(4, i), for i=1 to n, so the program uses two-dimensional arrays.

Add beneath the DIM statement in NPHASE

```
DIM xcc(4, n)
```

Then replace the initial conditions

```
xc(1) = 5# : xc(2) = 5# : xc(3) = 5#
```

with

```
xcc(1, 1) = 5.000# : xcc(1, 2) = 5# : xcc(1, 3) = 5#
xcc(2, 1) = 5.005# : xcc(2, 2) = 5# : xcc(2, 3) = 5#
xcc(3, 1) = 5.010# : xcc(3, 2) = 5# : xcc(3, 3) = 5#
xcc(4, 1) = 5.015# : xcc(4, 2) = 5# : xcc(4, 3) = 5#
```

Finally, delete the original PSET line, and replace the line GOSUB Runge with

```
FOR j = 1 TO 4
  FOR i = 1 TO n : xc(i) = xcc(j, i) : NEXT
    GOSUB Runge
  FOR i = 1 TO n : xcc(j, i) = xc(i) : NEXT
    PSET (xcc(j, 1), xcc(j, 3)), 10 + j
NEXT
```

If we run the original NPHASE, amended in this way, and INPUT r=28, we see a dramatic demonstration of sensitivity to initial conditions in the Lorenz equations (cf. Fig. 11.9). The values xi printed on screen at the end are those of the last-updated n-vector xcc(4, i), for i = 1 to n.

It can be instructive to watch the phase points move about *without* plotting the phase paths, and one way of doing this is to add

```
radius = xm/100
```

after, say, the INPUT command, then add

```
CLS
```

immediately after the DO command, and finally replace the PSET line by

```
CIRCLE (xcc(j, 1), xcc(j, 3)), radius, 10 + j
```

Program 5: NXT

```
DEFDBL A-H, K-M, O-Z: DEFINT I-J, N
n = 3: OPTION BASE 1
DIM x(n), xc(n), f(n), c1(n), c2(n), c3(n), c4(n)

REM ****** Setting up graphics ******

  CLS : SCREEN 9: PAINT (1, 1), 9
   xm = 4: tm = 50#
  VIEW (20, 30) - (575, 240), 0, 13
  WINDOW (0, -xm) - (tm, xm)
  LINE (0, 0) - (tm, 0), 11: LINE (0, -xm) - (0, xm), 11
     FOR d = .1 TO 1 STEP .1
     LINE (d * tm, 0) - (d * tm, xm / 40), 11
     NEXT
```

```
    LOCATE 2, 1: PRINT xm: LOCATE 10, 74: PRINT tm
    LOCATE 19, 43: PRINT "xi"
    LOCATE 22, 2: PRINT "Time"

REM ****** Step-by-step method ******

DO

  k = .05#: w = 1#: a = 7.5#
  LOCATE 19, 1
   INPUT "x1,x2,h,col"; xc(1), xc(2), h, col

  t = 0#: xc(3) = 0#

    DO
      GOSUB Runge
      t = t + h
      PSET (t, xc(1)), col
    LOOP UNTIL ABS(t - tm) < h / 2 OR INKEY$ = "q"
      LOCATE 22, 6: PRINT t
       FOR i = 1 TO n
        LOCATE 18 + i, 46: PRINT ; xc(i)
       NEXT

LOOP

REM ****** Subroutines ******

Equations:

    f(1) = x(2)
    f(2) = -k * x(2) - x(1) ^ 3# + a * COS(w * x(3))
    f(3) = 1#

    RETURN

Runge:   [as for NPHASE]
```

This is for the same kind of system as NPHASE, but it produces a graph of one of the variables against time t.

We are asked to INPUT initial conditions, the step size h, and a colour for the graph (col, an integer between 1 and 15). As soon as the graph is drawn, the outer `DO...LOOP` prompts us to repeat this procedure, while leaving the original graph on screen. In this way we can compare directly on screen the results from (a) different values of the step size h or (b) different initial conditions.

As it stands, the program is for the forced cubic oscillator equation (11.1), converted into an autonomous third-order system. It can be used to produce Fig. 11.1.

A variation: `NXTWAIT`

It can happen that we wish to give a system a chance to 'settle down' before we plot one of its variables against time t, and this simple variation on `NXT` allows us to do this. Replace the second line of 'Setting up graphics' by, say,

```
xm = 4 : twait = 50# : tm = 50#
```

replace the `PSET` command by

```
IF t > twait THEN
PSET (t - twait, xc(1)), col
END IF
```

and replace the `LOOP UNTIL...` line by

```
LOOP UNTIL ABS(t - twait - tm) < h/2 OR INKEY$ = "q"
```

The program will then plot `xc(1)` against `t` during the time interval `twait < t < twait + tm`.

Program 6: `NVARY`

```
DEFDBL A-H, K-M, O-Z: DEFINT I-J, N
n = 3: OPTION BASE 1
DIM x(n), xc(n), f(n), c1(n), c2(n), c3(n), c4(n)
DIM xcold(n)

REM ****** Setting up graphics ******

 CLS : SCREEN 9: PAINT (1, 1), 1
  xm = 8: ym = 8
 VIEW (180, 17) - (595, 330), 0, 14
 WINDOW (-xm, -ym) - (xm, ym)
 LINE (-xm, 0) - (xm, 0), 11
 LINE (0, -ym) - (0, ym), 11
```

```
  LOCATE 13, 76: PRINT xm: LOCATE 1, 48: PRINT ym
  LOCATE 15, 1: PRINT "xi"
  LOCATE 23, 2: PRINT "Time"

REM ****** Step-by-step method ******

 LOCATE 13, 2: INPUT "w"; w

 k = .1#: b = .04#

 xc(1) = 0#: xc(2) = 0#: xc(3) = 0#
 h = .1#

 DO

   FOR i = 1 TO n: xcold(i) = xc(i): NEXT
         CLS
         LINE (-xm, 0) - (xm, 0), 11
         LINE (0, -ym) - (0, ym), 11
         t = 0#

     DO
       GOSUB Runge
       t = t + h
       PSET (xc(1), xc(2)), 14
       LOCATE 23, 6: PRINT INT(t)
     LOOP UNTIL INKEY$ = "q"
      LOCATE 16, 1
       FOR i = 1 TO n: PRINT "                    ": NEXT
      LOCATE 16, 1
       FOR i = 1 TO n: PRINT CSNG(xc(i)): NEXT

   LOCATE 13, 2: INPUT "new w"; w
    IF w = 0# THEN
      FOR i = 1 TO n: xc(i) = xcold(i): NEXT
      LOCATE 13, 2: INPUT "new w"; w
    END IF

 LOOP

REM ****** Subroutines ******
```

```
Equations:

f(1) = x(2)
f(2) = -k * x(2) / w + (-x(1) - b * x(1) ^ 3#
       + COS(x(3))) / w ^ 2#      (as one line)
f(3) = 1#
RETURN

Runge:   [as for NPHASE]
```

This is another close relative of NPHASE, and is intended for investigating the effect of gradually varying a parameter which occurs in the differential equations.

As it stands, the program is for the forced Duffing equation (11.14), with the parameters as in Fig. 11.16. The drive frequency Ω—denoted by w in the program—is to be gradually varied. As a small change in Ω can produce an unintentionally large change in $\cos \Omega t$ if t has become large, we have rescaled the time variable to $t' = \Omega t$, so that (11.14) becomes

$$\frac{\mathrm{d}^2 x}{\mathrm{d}t'^2} + \frac{k}{\Omega}\frac{\mathrm{d}x}{\mathrm{d}t'} = \frac{-\alpha x - \beta x^3 + A\cos t'}{\Omega^2},$$

and this is the form of the equation in the program as it stands. It is probably simplest to use this particular example to show how the program works.

Run the program, INPUT w=1, and press q for 'quit' when the system has clearly settled into a periodic oscillation, corresponding to a closed path on screen. You will then be prompted to INPUT a new value of w. Set w=1.1, say, and the numerical integration will then *continue* in time from where it left off, taking the most recent values of xc(i), i=1,...,n as initial values for the subsequent calculation. Again, press q when the solution has settled down to a new, slightly different, periodic orbit.

If we continue doing this, in steps of 0.1, all goes smoothly until we change w from 1.5 to 1.6, at which stage the system suddenly collapses into a much smaller periodic oscillation. Identifying *sudden* or 'catastrophic' changes of this kind as some parameter is varied is mainly what NVARY is intended for.

At this point, of course, we would really like to increase w from 1.5 to 1.6 more gradually, both to confirm that there is, indeed, a 'jump', and to find more precisely where it occurs. To do this, after quitting the w=1.6 run, simply press RETURN (or ENTER). The computer interprets this as inputting w=0, and this triggers the contents of an IF...THEN...END IF statement which set the variables xc(i) back to their values just *before* the latest, eventful, run. If we now INPUT w=1.5 we recover the original oscillation just before the jump occurred, and can proceed to increase w from 1.5 again, but in smaller steps. By this technique we find that the jump to the smaller-amplitude oscillation takes place at w ≑ 1.52, and if we gradually *decrease* w again, in moderate steps at first if desired, we find a jump back to the larger-amplitude oscillation when w falls below 1.25 (Fig. 11.16).

More generally, NVARY is useful for exploring parameter space for a given set of differential equations and looking for interesting system behaviour. If we wish to be able to recapture that behaviour quickly on some later occasion, we need only note down the parameter value and the current values of xc(i), for i=1,...,n, displayed on screen. By using those xc(i) values as *initial conditions* when the program is started up again, we should be able to recapture immediately the system behaviour which caught our interest, and so continue the investigation.

Program 7: PENDANIM

```
DEFDBL A-H, K-M, O-Z: DEFINT I-J, N
n = 2: OPTION BASE 1
DIM x(n), xc(n), f(n), c1(n), c2(n), c3(n), c4(n)

REM ****** Setting up graphics ******

  CLS : SCREEN 9
  PAINT (1, 1), 9
  VIEW (180, 17) - (595, 330), 0, 14
  WINDOW (-2, -2) - (2, 2)
   LOCATE 22, 1: PRINT "Time"

REM ****** Step-by-step method ******

DO

    k = .1#
    t = 0#
    LOCATE 14, 1: INPUT "angle"; xc(1)
    LOCATE 15, 1: INPUT "angvel"; xc(2)
    h = .05#: animsteps = 4

     DO
        CLS
         b1 = SIN(xc(1)): b2 = -COS(xc(1))
         LINE (0, 0) - (b1, b2), 4
         CIRCLE (b1, b2), .05, 9: PAINT (b1, b2), 9
          FOR j = 1 TO animsteps
            GOSUB Runge
            t = t + h
          NEXT
        LOCATE 22, 6: PRINT CSNG(t)
```

```
    LOOP UNTIL INKEY$ = "q"

LOOP

REM ****** Subroutines ******

Equations:

 f(1) = x(2)
 f(2) = -k * x(2) - SIN(x(1))
 RETURN

Runge:   [as for NPHASE]
```

This is another variation on the key program NPHASE, for the damped simple pendulum equation $\ddot{\theta} + \tilde{k}\dot{\theta} + \sin\theta = 0$ (see Ex. 5.5), which we recast as

$$\dot{\theta} = y,$$

$$\dot{y} = -\tilde{k}y - \sin\theta$$

and then code with `x(1)`$=\theta$, `x(2)`$=y$. The novel feature here is the way in which the outcome is displayed as an *animation* of the pendulum motion.

Our method for doing this is simple but crude: each time the inner `DO...LOOP` is executed the graphics window is cleared (`CLS`), a line is drawn from the origin to the current position (b_1, b_2) of the pendulum bob, and the bob itself is drawn in and painted (`CIRCLE/PAINT`). This pendulum picture remains on screen while the computer performs a small number (`animsteps`) of up-datings using the Runge–Kutta method, and it is then wiped out and replaced by a new one at the next passage through the `DO...LOOP`. Adjusting the parameter `animsteps` can help keep animation flicker to a minimum.

As written, the program asks us to `INPUT "angle"`, i.e., the initial angle θ measured from the downward vertical in radians, and `"angvel"`, the initial dimensionless angular velocity $y = \dot{\theta}$, denoted by $\tilde{\Omega}$ in Section 5.5 and Ex. 5.5. When we press `q` for 'quit' the computation stops and we are prompted to enter new initial conditions.

With a few simple modifications `PENDANIM` can be used to illustrate a number of important concepts in Chapter 10:

Exercise 10.1: An elastic instability

Add just before the other `INPUT` commands

```
LOCATE 13, 1 : INPUT "S"; S
```

and add after the program lines which draw the pendulum

```
LINE (0, 1) - (b1, b2), 2
```

Finally, amend `Equations` in accord with Ex. 10.1, so that

```
f(2) = -k * x(2) - sin(x(1))
       + S * (2# * cos(x(1)/2#) - 1#) * sin(x(1)/2#)
```

The system in Fig. 10.5 may be explored quite well by having `k=.1#` and inputting the following sets of values for `S`, `angle`, and `angvel`: `(0,1.5,0)`, `(1.9,1.5,0)`, `(2.1,1.5,0)`, `(2.1,1.8,0)`, `(5,.0001,0)`, `(25,1E-20,0)`, the angle in the last set denoting 10^{-20}.

Exercise 10.3: Multiple equilibria and catastrophe

Begin by amending `Equations` in `PENDANIM` so that

```
f(2) = -k * x(2) + x(1) - x(1) ^ 3#/6# - (x(1) - eps)/m
```

which is the damped equivalent of (10.22).

The main change here, however, is so that after pressing `q` we can `INPUT` a new value of the parameter `eps` *without* going back to the initial conditions at $t=0$ again. We therefore take these out of the outer `DO...LOOP`, and replace the first six lines of the step-by-step method (including the first, but not second, `DO` statement) by

```
   t = 0#
   xc(1) = -.1# : xc(2) = 0#
 DO
   k = .1# : m = 1.2#
     IF m > 1 THEN
       critval = SQR(8 * (m - 1) ^ 3/(9 * m))
       LOCATE 5, 1 : PRINT CSNG (critval)
     END IF
   LOCATE 14, 1 : INPUT "eps"; eps
   c = .3 * cos(eps) : s = .3 * sin(eps)
   h = .05# : animsteps = 4
```

The `IF...END IF` section displays the critical value of ε given in Ex. 10.3 on screen, for convenience.

Finally, in the animation part of the inner `DO...LOOP`, replace `b2=-cos(xc(1))` by

```
b2 = cos(xc(1)),
```

because θ (i.e. xc(1)) is now measured from the *upward* vertical, and add after the CIRCLE/PAINT commands

```
LINE (0, 0) - (s, c), 1
LINE (-c, s) - (c, -s), 1
```

to provide a crude picture of the orientation of the board-and-springs arrangement in Fig. 10.11, which will change as we gradually vary eps.

With M=1.2, as given, try inputting eps=0, so that the rod flops down to the left-hand equilibrium, and then gradually increase eps in steps of, say, 0.02. When the rod suddenly 'jumps' to the right-hand equilibrium gradually reduce eps again in the same way.

Experimenting with different damping coefficients k and different M can be instructive, though if M is large the angles θ can be rather too large for (10.22) to be a good approximation of (10.19).

Finally, try changing the program so that eps is specified as, say, .1#, while M is gradually varied through the INPUT command. Start by inputting, say, M=2.5, and then gradually reduce M in steps of .2. A jump should occur, and gradually increasing M again in steps of .2 should lead smoothly to the *other* equilibrium position when M=2.5. Check that all this is consistent with Fig. 10.13.

Exercise 10.4: Hysteresis involving a state of motion

Take the program PENDANIM and amend Equations so that

```
f(2) = -k * x(2) - sin(x(1)) + torque
```

where torque is our coding for the dimensionless torque $\tilde{\Gamma}$. Then replace the first six lines of the step-by-step method by

```
  t = 0#
  xc(1) = 0# : xc(2) = 0#
DO
  k = .3#
  LOCATE 14, 1 : INPUT "torque"; torque
  h = .05# : animsteps = 4
```

so that the new program works in a similar way to that for Ex. 10.3.

On running the program, INPUT torque= .1, say, and press q for 'quit' when the pendulum has settled reasonably well into its equilibrium state. Increase torque to, say, .3, and continue in this way until the torque is 1.1, at which stage the pendulum should lock into a whirling motion. Then gradually *decrease* the torque again. With k=.3# the whirling motion only collapses when torque $<$ 0.37. With a larger value of the friction constant k=.8# the 'overlap' region where both equilibrium and whirling solutions exist is less marked, but still there: 0.86 $<$ torque $<$ 1. The nature of the whirling motion is particularly interesting just before it collapses, i.e., as we reduce the torque close to .86; the pendulum then spends a notable amount of each cycle almost motionless, and close to the *unstable* equilibrium position with $\cos \theta_0 < 0$.

A variation: VIBRAPEN

In this variation on PENDANIM the pivot is made to vibrate up and down with a prescribed amplitude and frequency. The system is governed by (12.9), or its non-dimensional equivalent (12.11), which may be expressed as the first-order system (12.13). As usual, the parameters occurring in the program, a, w, and k denote the *dimensionless* parameters of the model, $\tilde{a}$, $\tilde{\omega}$, and $\tilde{k}$, given by (12.12).

Begin by changing the original PENDANIM to

```
n = 3
```

and amending Equations to

```
f(1) = x(2)
f(2) = -k * x(2) - (1# + c * cos(w * x(3))) * sin(x(1))
f(3) = 1#
```

Then change the line t=0# to

```
t = 0# : xc(3) = 0#
```

and insert before the INPUT lines

```
LOCATE 13,1 : INPUT "a,w"; a,w
c = a * w ^ 2#
```

Next, try

```
h = .02# : animsteps = 2
```

though the value of animsteps for a reasonably smooth, pleasing animation will depend on the time step h and the graphics facilities on the computer being used.

Finally, replace the two lines

```
b1 = sin(xc(1)) : b2 = -cos(xc(1))
LINE (0, 0) - (b1, b2), 4
```

with

```
LINE (0, -a) - (0, a), 8
piv = -a * cos(w * xc(3))
CIRCLE (0, piv), .02, 12 : PAINT (0, piv), 12
b1 = sin(xc(1)) : b2 = -cos(xc(1)) + piv
LINE (0, piv) - (b1, b2), 4
```

One intriguing application of the above program is to the inverted pendulum of Ex. 12.1. In exploring the 'dancing' oscillations, and checking, in particular, whether they have settled down to a periodic limit-cycle form, as in Fig. 12.3, it can be helpful to replace the `CLS` statement inside the inner `DO...LOOP` by, say,

```
IF t < 70 THEN
CLS
END IF
```

Program 8: `PENDOUBL`

```
DEFDBL A-H, K-M, O-Z: DEFINT I-J, N
n = 5: OPTION BASE 1
DIM x(n), xc(n), f(n), c1(n), c2(n), c3(n), c4(n)

REM ****** Setting up graphics ******

  CLS : SCREEN 9
  PAINT (1, 1), 9
  VIEW (180, 17) - (595, 330), 0, 14
  WINDOW (-2.4, -2.4) - (2.4, 2.4)
   LOCATE 21, 2: PRINT "Time"

REM ****** Step-by-step method *******

DO

  k = .1#: m = .1#
```

```
  LOCATE 12, 2: INPUT "a,w"; a, w

  t = 0#: xc(2) = 0#: xc(4) = 0#: xc(5) = 0#
  LOCATE 14, 2: INPUT "ang1,ang2"; xc(1), xc(3)
  h = .05#

   DO
     CLS
      LINE (0, -a) - (0, a), 8
      ph = -a * COS(w * x(5))
      CIRCLE (0, ph), .025, 12: PAINT (0, ph), 12
     X1 = SIN(xc(1)): X2 = -COS(xc(1)) + ph
     X3 = SIN(xc(1)) + SIN(xc(3))
     X4 = -COS(xc(1)) - COS(xc(3)) + ph
      LINE (0, ph) - (X1, X2), 12
       CIRCLE (X1, X2), .05, 9: PAINT (X1, X2), 9
      LINE (X1, X2) - (X3, X4), 12
       CIRCLE (X3, X4), .05, 9: PAINT (X3, X4), 9

       GOSUB Runge
       t = t + h
     LOCATE 21, 6: PRINT CSNG(t)
   LOOP UNTIL INKEY$ = "q"
LOOP

REM ****** Subroutines ******

Equations:

  Q = x(3) - x(1): c = COS(Q): s = SIN(Q): P = c * s
  D = 1# - m * c ^ 2#
  g = (1# + a * w ^ 2# * COS(w * x(5)))
  g1 = g * SIN(x(1))
  g3 = g * SIN(x(3))
  x22 = x(2) ^ 2#: x42 = x(4) ^ 2#

  f(1) = x(2)
  f(2) = -k * x(2)
         + (m * (x42 * s + x22 * P + g3 * c) - g1) / D
  f(3) = x(4)
  f(4) = -k * x(4)
         + (-x22 * s - m * x42 * P - g3 + g1 * c) / D
  f(5) = 1#
  RETURN

Runge:   [as for NPHASE]
```

This program provides an animation of a *double* pendulum which has its pivot vibrating up and down with prescribed amplitude a and frequency ω (Sections 12.4, 12.5). It is therefore a two-pendulum counterpart to the program VIBRAPEN.

The basic equations are obtained by first replacing g by $g + a\omega^2 \cos \omega t$ in (12.15). The resulting equations are then non-dimensionalized by the transformation $\tilde{t} = (g/l)^{1/2}t$ and solved as simultaneous equations for $\ddot{\theta}_1$ and $\ddot{\theta}_2$, where a dot denotes differentiation with respect to $\tilde{t}$. Next, damping is introduced very crudely by replacing $\ddot{\theta}_1$ by $\ddot{\theta}_1 + \tilde{k}\dot{\theta}_1$ and treating $\ddot{\theta}_2$ likewise (cf. (12.11)). Finally, the equations are converted in the usual way into a fifth-order autonomous system in the variables x(1) $= \theta_1$, x(2) $= \dot{\theta}_1$, x(3) $= \theta_2$, x(4) $= \dot{\theta}_2$, x(5) $= \tilde{t}$.

The parameters are as in (12.12), with l denoting the length of *each* pendulum involved, and there is an additional parameter $m = m_2/(m_1 + m_2)$ (see (12.3)). As it stands, the program has $\tilde{k} = 0.1$, $m = 0.1$, and we INPUT values for the *dimensionless* drive amplitude $\tilde{a}$ and frequency $\tilde{\omega}$. It also sets the initial angular velocities $\dot{\theta}_1$, $\dot{\theta}_2$ to zero, and calls for us to INPUT the initial angles, measured (in radians) from the downward vertical, as in Fig. 12.6.

If the drive frequency $\tilde{\omega}$ is large—as it can be when stabilizing the system in its *inverted* state—it may be necessary to make the time step h quite small in order to obtain consistent results which pass the 'halve the step-size' test. Whether we have this consistency or not cannot really be seen from the animation, but we can modify the program slightly so that it runs for a specified time tm and then prints, say, xc(1) and xc(3) at the end of that time.

Program 9: THREEBP

```
DEFDBL A-H, K-M, O-Z: DEFINT I-J, N
n = 12: OPTION BASE 1
DIM x(n), xc(n), f(n), c1(n), c2(n), c3(n), c4(n)

REM ****** Setting up graphics ******

  CLS : SCREEN 9
  PAINT (1, 1), 9
   xm = .75: ym = xm
  VIEW (180, 17) - (595, 330), 0, 13
  WINDOW (-xm, -ym) - (xm, ym)
  LINE (-xm, 0) - (xm, 0), 8: LINE (0, -ym) - (0, ym), 8
  LOCATE 13, 76: PRINT xm
```

```
REM ****** Step-by-step method ******

  m1 = .5#: m2 = .5#: m3 = .5#
  LOCATE 13, 1: INPUT "x3,y3"; x3, y3

  t = 0#

     REM ** initial x-coordinates **
   xc(1) = -.5#: xc(2) = .5#: xc(3) = x3
     REM ** initial y-coordinates **
   xc(4) = 0#: xc(5) = 0#: xc(6) = y3
     REM ** initial x-velocities **
   xc(7) = 0#: xc(8) = 0#: xc(9) = 0#
     REM ** initial y-velocities **
   xc(10) = -.3#: xc(11) = .3#: xc(12) = -.3#

  h = .003#

  DO
    GOSUB Runge
    PSET (xc(1), xc(4)), 12
    PSET (xc(2), xc(5)), 9
    PSET (xc(3), xc(6)), 15
    t = t + h

     REM ** Energy conserved? **
     kin1 = .5# * m1 * (xc(7)  ^  2# + xc(10)  ^  2#)
     kin2 = .5# * m2 * (xc(8)  ^  2# + xc(11)  ^  2#)
     kin3 = .5# * m3 * (xc(9)  ^  2# + xc(12)  ^  2#)
     pot = -(m1 * m2 / r12 + m2 * m3 / r23
           + m3 * m1 / r31)
     energy = kin1 + kin2 + kin3 + pot
     LOCATE 21, 1: PRINT "Energy"
     LOCATE 22, 1: PRINT energy

    LOCATE 17, 7: PRINT "                "
    LOCATE 17, 1: PRINT "Time ="; CSNG(t)

  LOOP UNTIL INKEY$ = "q"

END

REM ****** Subroutines ******
```

```
Equations:

 d21 = x(2) - x(1): d32 = x(3) - x(2): d13 = x(1) - x(3)
 d54 = x(5) - x(4): d65 = x(6) - x(5): d46 = x(4) - x(6)

 r12 = (d21 ^ 2# + d54 ^ 2#) ^ .5#
 r23 = (d32 ^ 2# + d65 ^ 2#) ^ .5#
 r31 = (d13 ^ 2# + d46 ^ 2#) ^ .5#

 p12 = r12 ^ 3#: p23 = r23 ^ 3#: p31 = r31 ^ 3#

 f(1) = x(7): f(2) = x(8): f(3) = x(9)
 f(4) = x(10): f(5) = x(11): f(6) = x(12)

 f(7) = m2 * d21 / p12 - m3 * d13 / p31
 f(8) = m3 * d32 / p23 - m1 * d21 / p12
 f(9) = m1 * d13 / p31 - m2 * d32 / p23
 f(10) = m2 * d54 / p12 - m3 * d46 / p31
 f(11) = m3 * d65 / p23 - m1 * d54 / p12
 f(12) = m1 * d46 / p31 - m2 * d65 / p23

 RETURN

Runge:  [as for NPHASE]
```

This program is a fairly straightforward variation on `NPHASE` to integrate the system of 12 equations of first order which correspond to the three-body problem (6.49). Those equations are entered into the program as follows:

Dimensionless variable			Variable in subroutine 'Equations'		
$\tilde{x}_1$	$\tilde{x}_2$	$\tilde{x}_3$	`x(1)`	`x(2)`	`x(3)`
$\tilde{y}_1$	$\tilde{y}_2$	$\tilde{y}_3$	`x(4)`	`x(5)`	`x(6)`
$\dot{\tilde{x}}_1$	$\dot{\tilde{x}}_2$	$\dot{\tilde{x}}_3$	`x(7)`	`x(8)`	`x(9)`
$\dot{\tilde{y}}_1$	$\dot{\tilde{y}}_2$	$\dot{\tilde{y}}_3$	`x(10)`	`x(11)`	`x(12)`

though the actual quantities which are continually updated and plotted are denoted, as elsewhere, by `xc(1)`, etc. The constants `m1`, `m2`, `m3` in the program denote the dimensionless masses $\tilde{m}_1$, $\tilde{m}_2$, $\tilde{m}_3$.

As it stands the program uses a constant time step h, but to implement the variable time step (6.50) we need only replace the statement `h=.003#` by, say,

```
hscale = .1#
```

and then add

```
h = hscale/(r12 ^ -2# + r23 ^ -2# + r31 ^ -2#)
```

just before the `LOOP...` statement.

The masses and initial conditions are set up for the example of Fig. 6.15. To work through the sequence in the text, start by changing `m3` to, say, `.00005#` and the initial values of `xc(10)` and `xc(11)` to `-.5#` and `.5#`, respectively. On running the program, `INPUT`, say, `x3=50`, `y3=50`, to start m_3 well away from m_1 and m_2. The masses m_1 and m_2 will then move in a circle of radius 0.5 about the origin. Next, change `xc(10)` and `xc(11)` back to their original values of `-.3#` and `.3#`, repeat the `INPUT` as before, and obtain the elliptical orbits of Fig. 6.15(a).

Now set `m3=.5#`, and `INPUT` `x3=-0.1`, `y3=0.75`. With a fixed step size `h=.003` we can then reproduce the sequence in Fig. 6.15, but only just, because of the build-up of error resulting from the sparse distribution of successive points during 'close encounters'. By $\tilde{t} \approx 4.8$ the total energy has begun to change significantly, and the whole step-by-step method goes wildly wrong a short time later ($\tilde{t} \approx 5.7$) during a close encounter between m_1 and m_2. Decreasing h to 0.001 or even 0.0003 does not resolve the problem, but we can perform a longer time integration successfully by using a variable step length instead with, say, `hscale=0.1` or `0.05`.

In the case of Fig. 6.16 a fixed step size `h=.003` will, again, produce a broadly correct sequence of events until the very close encounter between m_1 and m_2 at $\tilde{t} = 23.56$, but the outcome of that encounter is then completely wrong. This will happen even with a variable time step if `hscale` is too large, say `0.2` or more. We apparently obtain consistent results, however, with `hscale` equal to `0.1`, or `0.05`, or `0.025`.*

One way of checking for such consistency is to use the method in `NXT` to get two 'solutions' on screen at once. Thus, delete the line

```
hscale = .1#
```

* The computing time for the last-mentioned calculation may be quite long, perhaps half-an-hour or so on a 486 DX2 66 using interpreted QBasic, though only five minutes or so on a fast Pentium.

change the colours on the `PSET` statements from `12`, `9`, `15` to `k1`, `k2`, `k3`, add after the `INPUT` command

```
LOCATE 14, 1 : INPUT "hscale"; hscale
LOCATE 15, 1 : INPUT "k1,k2,k3"; k1,k2,k3
```

and put an outer `DO...LOOP` around the whole of the step-by-step method.

Program 10: `HEAT`

```
REM ****** Setting up graphics ******

  CLS : SCREEN 9: PAINT (1, 1), 1
   xm = 1: ym = .5: tm = .2
  VIEW (20, 30) - (575, 240), 0, 11
  WINDOW (0, 0) - (xm, ym)
   LOCATE 18, 74: PRINT xm
   LOCATE 2, 2: PRINT ym
   LOCATE 23, 1: PRINT "y(middle)="
   LOCATE 23, 30: PRINT "t="

REM ****** Finite-difference grid parameters ******

  LOCATE 19, 1: INPUT "m"; m
    DIM y(m), ynew(m)
    h = xm / m
    LOCATE 19, 10: PRINT "h"; h
    LOCATE 20, 1: PRINT "kcrit="; .5 * h ^ 2
  LOCATE 21, 1: INPUT "k < kcrit"; k

REM ****** Initial conditions ******

  DEF fnf (x) = .5 * EXP(-100 * (x - .5) ^ 2)

    FOR i = 1 TO m - 1
      y(i) = fnf(i * h)
    NEXT
      y(0) = 0: y(m) = 0

REM ****** Step-by-step method ******

    t = 0
```

```
DO

   CLS

     LINE (0, 0) - (h, y(1)), 9
   FOR i = 1 TO m - 1
     LINE (i * h, y(i)) - ((i + 1) * h, y(i + 1)), 9
     diff2 = y(i + 1) - 2 * y(i) + y(i - 1)
     ynew(i) = y(i) + k * (diff2 / h ^ 2)
   NEXT
     ynew(0) = 0: ynew(m) = 0

   FOR i = 0 TO m
     y(i) = ynew(i)
   NEXT

   t = t + k

    LOCATE 23, 11: PRINT INT(y(m / 2) * 1000) / 1000
    LOCATE 23, 32: PRINT INT(t * 1000) / 1000

LOOP UNTIL ABS(t - tm) < k / 2 OR INKEY$ = "q"
```

This program integrates the dimensionless heat equation (7.16) subject to the boundary conditions (7.17) and initial condition (7.18). The dependent variable is denoted by `y` rather than T.

The program proceeds by using the values of y at the grid points of one horizontal row in Fig. 7.8 (`y(i)`, `i=0` to `m`) to determine the unknown values of `y` in the next row up (`ynew(i)`, `i=1` to `m-1`), according to the rule (7.23), which is divided between two lines in the program itself. These 'new' values `ynew(i)` then become the 'old' ones `y(i)` for the next cycle through the `DO...LOOP`. At each stage, the variables `y(i)` are plotted, not by using the `PSET` command but by using the `LINE` command to 'join up the dots' and produce a continuous curve.

As written, the program has an initial state $y = f(\tilde{x})$ consisting of a localized 'hot spot' centred on $\tilde{x} = 0.5$. If we `INPUT`, say, `m=50`, the program then causes the largest time step `kcrit` consistent with the stability criterion (7.24) to appear on screen. On inputting, say, `k=.0001`, we see the hot spot diffuse, quickly at first and more slowly later on (cf. Fig. 7.7).

Inputting a value of `k` slightly *greater* then `kcrit` will lead to an interesting and dramatic numerical instability. This is worth seeing, but the author's experience is that it can 'crash' the computer system, so it is worth being ready to press `q` or `CTRL-Break` as soon as the oscillations begin to get out of hand.

Solutions to the exercises

For exercises where some computation is required, the appropriate sections of Appendices A and B should also be consulted.

Chapter 2

2.1
$$\frac{\mathrm{d}y}{\mathrm{d}x} = 3x^2 - a, \qquad \frac{\mathrm{d}^2y}{\mathrm{d}x^2} = 6x.$$

If $a < 0$ there are no stationary points. If $a > 0$ there are two, at $x = -(a/3)^{1/2}$ and $x = +(a/3)^{1/2}$. The first is a local maximum, because $\mathrm{d}^2y/\mathrm{d}x^2 < 0$ there, so the slope of the curve is decreasing with x at that point. The second is a local minimum.

If $a = 0$ there is a stationary point at $x = 0$, but it is neither a local maximum nor a local minimum.

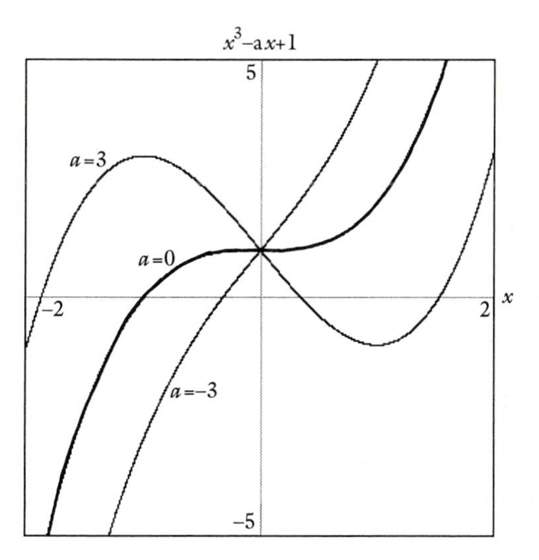

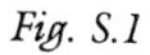

Fig. S.1

2.2 By (2.7b),

$$\frac{\mathrm{d}}{\mathrm{d}x}\left\{\frac{\exp(x+y)}{\exp(x)}\right\} = \frac{1}{[\exp(x)]^2}\left\{\exp(x)\frac{\mathrm{d}}{\mathrm{d}x}\exp(x+y) - \exp(x+y)\frac{\mathrm{d}}{\mathrm{d}x}\exp(x)\right\}.$$

On writing $u = x + y$ and using the chain rule (2.9) we see that

$$\frac{\mathrm{d}}{\mathrm{d}x}(\exp u) = \frac{\mathrm{d}}{\mathrm{d}u}(\exp u)\cdot\frac{\mathrm{d}u}{\mathrm{d}x} = \exp(u),$$

in view of (2.20) and the fact that y is being kept constant. The term in curly brackets therefore reduces to

$$\exp(x)\exp(x+y) - \exp(x+y)\exp(x)$$

where we have again used (2.20). So

$$\frac{\mathrm{d}}{\mathrm{d}x}\left\{\frac{\exp(x+y)}{\exp(x)}\right\} = 0,$$

and $\exp(x+y)/\exp(x)$ is therefore a constant. Using (2.21) we find that constant to be $\exp(y)$, which proves the result.

2.3
$$\begin{aligned}\frac{1}{x} &= \lim_{n\to\infty} n\left[\log\left(x+\frac{1}{n}\right) - \log x\right] \\ &= \lim_{n\to\infty} n\log\left(1+\frac{1}{nx}\right) \\ &= \lim_{n\to\infty} \log\left(1+\frac{1}{nx}\right)^n,\end{aligned}$$

using well-known properties of the function $\log x$ which may be deduced directly from (2.30). So

$$\lim_{n\to\infty}\left(1+\frac{1}{nx}\right)^n = \mathrm{e}^{1/x},$$

and on setting $x = 1/\alpha$ the result follows.

2.4 The derivative of e^{-x} is $-\mathrm{e}^{-x}$, by (2.27), so

$$\frac{\mathrm{d}}{\mathrm{d}x}\cosh x = \tfrac{1}{2}(\mathrm{e}^x - \mathrm{e}^{-x}) = \sinh x,$$

and the second result follows likewise.

$$\cosh^2 x - \sinh^2 x = \tfrac{1}{4}(e^{2x} + 2 + e^{-2x}) - \tfrac{1}{4}(e^{2x} - 2 + e^{-2x}) = 1.$$

2.5
$$\begin{aligned} e^{i\theta_1} e^{i\theta_2} &= (\cos\theta_1 + i\sin\theta_1)(\cos\theta_2 + i\sin\theta_2) \\ &= [\cos\theta_1\cos\theta_2 - \sin\theta_1\sin\theta_2 \\ &\quad + i\{\sin\theta_1\cos\theta_2 + \cos\theta_1\sin\theta_2\}] \\ &= \cos(\theta_1+\theta_2) + i\sin(\theta_1+\theta_2) \\ &= e^{i(\theta_1+\theta_2)}. \end{aligned}$$

Chapter 3

3.1 The integrating factor is $e^{\int 2t\,dt} = e^{t^2}$,

so
$$\frac{d}{dt}(x\,e^{t^2}) = t\,e^{t^2},$$

$\therefore$
$$x\,e^{t^2} = \tfrac{1}{2}e^{t^2} + c,$$
$$x = \tfrac{1}{2} + c\,e^{-t^2}.$$

But $x = 1$ when $t = 0$, so $c = \frac{1}{2}$ and

$$x = \tfrac{1}{2}(1 + e^{-t^2}).$$

3.2 The equation is separable, so

$$\int \frac{dx}{x^2} = \int \frac{dt}{1+t}$$

$\therefore$
$$-\frac{1}{x} = \log(1+t) + c.$$

But $x = 1$ when $t = 0$, so $c = -1$ and

$$x = \frac{1}{1 - \log(1+t)}.$$

The solution 'blows up' when $\log(1+t) = 1$, i.e. when $t = e - 1$.

3.3 (a) Substituting (3.24) into the left-hand side of (3.23) gives

$$\begin{aligned} &a(A\ddot{x}_1 + B\ddot{x}_2) + b(A\dot{x}_1 + B\dot{x}_2) + c(Ax_1 + Bx_2) \\ &\quad = A(a\ddot{x}_1 + b\dot{x}_1 + cx_1) + B(a\ddot{x}_2 + b\dot{x}_2 + cx_2) = 0. \end{aligned}$$

Note that this idea of linearly combining two solutions to form another solution would *not* have worked if, say, the term cx in (3.23) had been cx^3, or if the equation had been otherwise nonlinear.

(b) Using (3.29) and (3.30) the general solution is

$$x = A\,\mathrm{e}^t + B\,\mathrm{e}^{-t}.$$

So $\dot{x} = A\,\mathrm{e}^t - B\,\mathrm{e}^{-t}$, and the initial conditions give $A + B = 1$, $A - B = 0$, so $A = B = \frac{1}{2}$. So

$$x = \tfrac{1}{2}(\mathrm{e}^t + \mathrm{e}^{-t}).$$

3.4 We have $a\ddot{x}_1 + b\dot{x}_1 + cx_1 = 0$. Writing $x = x_1 u$ we obtain

$$\dot{x} = \dot{x}_1 u + x_1 \dot{u}$$

$$\ddot{x} = \ddot{x}_1 u + 2\dot{x}_1 \dot{u} + x_1 \ddot{u}$$

so

$$(a\ddot{x}_1 + b\dot{x}_1 + cx_1)u + (2a\dot{x}_1 + bx_1)\dot{u} + ax_1\ddot{u} = 0.$$

The first bracket vanishes, and putting $z = \dot{u}$ we are left with

$$ax_1\dot{z} + (2a\dot{x}_1 + bx_1)z = 0,$$

i.e. a first-order equation for z.

In the case

$$\ddot{x} - 2\dot{x} + x = 0,$$

trying $x = \mathrm{e}^{mt}$ gives, from (3.31), $(m-1)^2 = 0$, producing just one solution $x_1 = \mathrm{e}^t$. As $a = 1$, $b = -2$, $c = 1$ the above method gives

$$\mathrm{e}^t\,\dot{z} + (2\,\mathrm{e}^t - 2\,\mathrm{e}^t)z = 0,$$

so $\dot{z} = 0$ and therefore z is a constant in this case, say B. But $\dot{u} = z$, so $\dot{u} = B$ and therefore $u = Bt + A$, say. The general solution in this case is therefore

$$x = (A + Bt)\,\mathrm{e}^t,$$

and on applying the initial conditions we find

$$x = (1 - t)\,\mathrm{e}^t.$$

3.5 Begin by spotting that $x = -t$ is a particular solution $x_{\mathrm{p}}(t)$, because the term $\ddot{x}$ is then zero. By (3.30) the general solution of the homogeneous equation is $A\,\mathrm{e}^t + B\,\mathrm{e}^{-t}$, so

$$x = A\,\mathrm{e}^t + B\,\mathrm{e}^{-t} - t.$$

Thus $\dot{x} = A\,\mathrm{e}^t - B\,\mathrm{e}^{-t} - 1$, and the initial conditions then give $A + B = 1$, $A - B - 1 = 0$, so $A = 1$ and $B = 0$ and

$$x = \mathrm{e}^t - t.$$

3.6 (3.44) gives

$$V(x)=\int_0^x \alpha s \, \mathrm{d}s=\tfrac{1}{2}\alpha x^2$$

in this case, so (3.48) becomes

$$\pm\int_{x_0}^{x}\frac{\mathrm{d}x}{\sqrt{x_0^2-x^2}}=\left(\frac{\alpha}{m}\right)^{1/2}t.$$

On introducing the change of variable $x=x_0\cos\theta$ we obtain

$$\mp\int_0^{\cos^{-1}(x/x_0)}d\theta=\left(\frac{\alpha}{m}\right)^{1/2}t,$$

i.e.

$$\cos^{-1}\frac{x}{x_0}=\mp\left(\frac{\alpha}{m}\right)^{1/2}t,$$

so that

$$x=x_0\cos\left(\frac{\alpha}{m}\right)^{1/2}t.$$

This solution could alternatively have been obtained by writing $\omega^2=\alpha/m$ in (3.27) and then using (3.28) and the initial conditions.

Chapter 4

4.1 Inserting

```
FOR t = 0 TO tm STEP tm/1000
  PSET (t, EXP(t)), 13
NEXT
```

after the `WINDOW` line is one way of plotting the 'actual' curve, in pink (`13`).
In Fig. 4.6(b) the smaller value of h only postpones the breakdown time, which appears to be roughly proportional to h^{-2}.

4.2 Euler's method applied to (4.28) gives

$$x_{n+1}=x_n+hy_n$$

$$y_{n+1}=y_n+h(-x_n),$$

but the alternative updating process given uses the just-updated value of x in the second line, not the original, so corresponds to

$$x_{n+1} = x_n + hy_n,$$

$$y_{n+1} = y_n + h(-x_{n+1}).$$

(The algorithm nonetheless works rather well, in fact, for the differential equation at hand, with *no* spurious growth in amplitude of the oscillations, even with fairly large h.)

4.3 A suitable improved Euler program for this case in which $f(x,t)=x$ is

```
h = .1 : tm = 1
t = 0 : x = 1
  DO
    c1 = h * (x)
    c2 = h * (x+c1)
    x = x + (c1 + c2)/2
    t = t + h
  LOOP UNTIL ABS(t - tm) < h/2
PRINT h, x, x - EXP(1)
```

On running it on my own PC I get

.1	2.7$\underline{1}$4081	$-4.201174\cdot 10^{-3}$
.01	2.718$\underline{2}$36	$-4.529953\cdot 10^{-5}$
.001	2.71828$\underline{1}$	$-4.768372\cdot 10^{-7}$

as a counterpart to Table 4.1, where the underlining shows the last correct digit.

A double-precision Runge–Kutta program is

```
DEFDBL C, H, T, X
h = .1# : tm = 1#
t = 0# : x = 1#
  DO
    c1 = h * (x)
    c2 = h * (x + c1/2#)
    c3= h * (x + c2/2#)
    c4 = h * (x + c3)
    x = x + (c1 + 2# * c2 + 2# * c3 + c4)/6#
    t = t + h
  LOOP UNTIL ABS(t - tm) < h/2#
PRINT h, x, x - EXP(1#)
```

On running it on my own PC I get

.1	$2.718\underline{2}79744135166$	$-2.084323879270045 \cdot 10^{-6}$
.01	$2.7182818\underline{2}8234403$	$-2.246416386242345 \cdot 10^{-10}$
.001	$2.7182818284590\underline{2}5$	$-2.042810365310288 \cdot 10^{-14}$

which is impressive evidence of the accuracy of the Runge–Kutta method. Note that error $\propto h^4$.

4.4
$$\begin{aligned}
\text{c1} &= h*(1) = h \\
\text{c2} &= h(1+\tfrac{1}{2}c_1) = h(1+\tfrac{1}{2}h) \\
\text{c3} &= h[1+\tfrac{1}{2}h(1+\tfrac{1}{2}h)] \\
\text{c4} &= h[1+h\{1+\tfrac{1}{2}h(1+\tfrac{1}{2}h)\}]
\end{aligned}$$

$$\begin{aligned}
x &= 1+\tfrac{1}{6}\big[h+2h(1+\tfrac{1}{2}h)+2h+h^2(1+\tfrac{1}{2}h) \\
&\quad +h+h^2\{1+\tfrac{1}{2}h(1+\tfrac{1}{2}h)\}\big] \\
&= 1+\tfrac{1}{6}\big[6h+3h^2+h^3+\tfrac{1}{4}h^4\big] \\
&= 1+h+\frac{h^2}{2!}+\frac{h^3}{3!}+\frac{h^4}{4!}
\end{aligned}$$

4.5 The Euler curve in Fig. 4.10 is not reliable, because if we rerun the calculation with $h=0.02$, say, rather than 0.035, we obtain a substantially different curve.

The Runge–Kutta curve *does* pass this 'halve the step size' test, however, even with values of h that are not particularly small. Moreover, it is always sound policy in any case of real doubt to check a computation using a *different numerical method*. In this case, the improved Euler method with $h=0.003$, say, gives the curve obtained by the Runge–Kutta method in Fig. 4.10, and even the Euler method itself can confirm it if we take, say, $h=0.00001$ and convert the program to double precision, to reduce the build-up of rounding errors from the very large number of steps.

4.6 From (2.14) we have

$$x(h) = x(0) + h\dot{x}(0) + \tfrac{1}{2}h^2\ddot{x}(0) + O(h^3).$$

The differential equation itself gives $\dot{x}=f(x)$, and $x=x_0$ at $t=0$, so $\dot{x}(0)=f(x_0)$. Also, differentiating the equation with respect to t and using the chain rule (2.9) gives

$$\ddot{x} = f'(x)\dot{x} = f'(x)f(x),$$

so $\ddot{x}(0)=f'(x_0)f(x_0)$, and the given result then follows.

One step of the improved Euler method in this autonomous case gives

$$c_1 = hf(x_0), \qquad c_2 = hf(x_0 + c_1)$$
$$x(h) = x_0 + \tfrac{1}{2}(c_1 + c_2),$$

i.e.

$$x(h) = x_0 + \tfrac{1}{2}h[f(x_0) + f(x_0 + c_1)].$$

But expanding $f(x_0 + c_1)$ as a Taylor series about x_0, recognizing that c_1 is small, of order h, gives

$$\begin{aligned} f(x_0 + c_1) &= f(x_0) + c_1 f'(x_0) + O(c_1^2) \\ &= f(x_0) + hf(x_0)f'(x_0) + O(h^2), \end{aligned}$$

so

$$x(h) = x_0 + hf(x_0) + \tfrac{1}{2}h^2 f(x_0)f'(x_0) + O(h^3).$$

The improved Euler method has an error of only $O(h^3)$, therefore, after just *one* step.

Chapter 5

5.1 Trying $x = \mathrm{e}^{mt}$ gives, according to (3.31),

$$m^2 + km + \omega^2 = 0,$$

so

$$m = -\frac{k}{2} \pm \left(\frac{k^2}{4} - \omega^2\right)^{1/2}.$$

As $\omega^2 > k^2/4$ we may write this more usefully as

$$m = -\frac{k}{2} \pm \mathrm{i}\left(\omega^2 - \frac{k^2}{4}\right)^{1/2}.$$

According to (3.32) the general solution is then

$$x = \mathrm{e}^{-kt/2}[E\,\mathrm{e}^{\mathrm{i}\theta} + F\,\mathrm{e}^{-\mathrm{i}\theta}]$$

where $\theta = (\omega^2 - \frac{1}{4}k^2)^{1/2}t$. On using (2.33) we may rewrite the quantity in brackets as $P\cos\theta + Q\sin\theta$, where $P = E + F$ and $Q = \mathrm{i}(E - F)$, and this in turn can be cast in the form $C\cos(\theta - D)$ by the same argument that leads from (5.13a) to (5.13b).

Casual use of 2XTPHASE shows that larger values of the parameter k lead to faster damping of the oscillations. The precise decay factor of $\mathrm{e}^{-kt/2}$ can be confirmed most simply, perhaps, by changing the first PSET command to

```
PSET(t, x * exp(.5 * k * t)), 13
```

for the x-t plot then displays the oscillations as having (apparently) constant amplitude.

5.2 Use the method at the end of Section 3.4. Trying a particular integral $x_p = C \cos \Omega t$ succeeds, with C turning out to be $a/(\omega^2 - \Omega^2)$. The complete solution is then

$$x = A \cos \omega t + B \sin \omega t + \frac{a}{\omega^2 - \Omega^2} \cos \Omega t,$$

and as $x = \dot{x} = 0$ at $t = 0$ we have $B = 0$ and $A = -a/(\omega^2 - \Omega^2)$, whence the result.

The above is for $\Omega \neq \omega$, of course. When $\Omega = \omega$ consider

$$x = \frac{a}{2\omega} t \sin \omega t,$$

as suggested. Then

$$\dot{x} = \frac{a}{2\omega}(\sin \omega t + \omega t \cos \omega t)$$

$$\ddot{x} = \frac{a}{2\omega}(2\omega \cos \omega t - \omega^2 t \sin \omega t)$$

so $\ddot{x} + \omega^2 x = a \cos \omega t$, as desired. Moreover, $x = \dot{x} = 0$ when $t = 0$.

5.3 (5.21) generalizes to

$$\left(2 - \frac{m_1}{\alpha} \omega^2\right) A = B,$$

$$A = \left(2 - \frac{m_2}{\alpha} \omega^2\right) B$$

and so

$$\left(2 - \frac{m_1}{\alpha} \omega^2\right)\left(2 - \frac{m_2}{\alpha} \omega^2\right) = 1$$

with roots

$$\frac{\omega^2}{\alpha} = \frac{m_1 + m_2 \pm \left[(m_1 + m_2)^2 - 3m_1 m_2\right]^{1/2}}{m_1 m_2}.$$

We may rewrite this as

$$m_1 \frac{\omega^2}{\alpha} = 1 + \frac{m_1}{m_2} \pm \left(1 - \frac{m_1}{m_2} + \frac{m_1^2}{m_2^2}\right)^{1/2},$$

and if $m_1/m_2 \ll 1$ then expanding the last term binomially (see (2.16)) gives, correct to first order in m_1/m_2,

$$m_1 \frac{\omega^2}{\alpha} \doteqdot 1 + \frac{m_1}{m_2} \pm \left(1 - \frac{m_1}{2m_2}\right)$$

$$\doteqdot 2 \quad \text{or} \quad \frac{3m_1}{2m_2}.$$

The first of these, with $\omega^2 \doteqdot 2\alpha/m_1$, is the fast mode, and the relation $A = (2 - m_2\omega^2/\alpha)B$ becomes approximately $B = -m_1 A/2m_2$. The mass m_2 therefore hardly moves in comparison with the first mass, and it is quite easy to see that $\omega^2 = 2\alpha/m_1$ is the *exact* result in the single-mass problem, with m_2 being held fixed.

The second, with $\omega^2 \doteqdot 3\alpha/2m_2$, is a very slow mode, and either expression relating A and B gives $B \doteqdot 2A$. This again makes sense; the two quite different masses are oscillating at the same low frequency, and the reason that the much smaller mass is not oscillating much faster is that with $B \doteqdot 2A$ the springs on either side of it are being stretched by *almost equal amounts*, so that their forces on that small mass are almost cancelling one another.

5.4 `xm = 4 : ym = 4 : tm = 50`

Delete `w = 1 : k = .1`

```
DEF fng (x, y, t) = -SIN(x)
```

At the `INPUT` prompt, make $x_0 = 3.124139$, i.e. 179°, and $y_0 = 0$.

One reasonably satisfactory way of obtaining a *half*-period, if $y_0 = 0$, is to replace the `LOOP` line by

```
LOOP UNTIL y > 0.
```

With `ImpEuler` and `h` in the region of 0.005 or so I consistently obtain

$$\theta_0 = 178^\circ \qquad T = 21.7$$
$$\theta_0 = 179^\circ \qquad T = 24.5$$
$$\theta_0 = 179.5^\circ \qquad T = 27.3$$

and have confirmed these values for the (dimensionless) oscillation period T by using `NXT` instead, which uses the Runge–Kutta method and double precision. Note that the small-amplitude oscillation period in these same dimensionless units is only $2\pi = 6.3$.

5.5 In dimensionless form, the problem is (5.46) with $\theta=0$, $y=\tilde{\Omega}$ at $\tilde{t}=0$. Proceeding as in Section 3.5 we obtain

$$\frac{d\theta}{dy}=\frac{y}{-\sin\theta},$$

which is separable, and implies that

$$-\int \sin\theta \, d\theta=\int y \, dy,$$

i.e.

$$\tfrac{1}{2}y^2=\cos\theta+\text{constant}.$$

On applying the initial conditions we find

$$\tfrac{1}{2}y^2=\cos\theta-1+\tfrac{1}{2}\tilde{\Omega}^2,$$

and as the left-hand side cannot be negative we find that $\cos\theta$ can only become equal to -1, corresponding to θ reaching π, if $\tilde{\Omega}^2\geq 4$, i.e. $\tilde{\Omega}\geq 2$. In view of (5.45) the result then follows.

Last part: $\tilde{\Omega}=4$ leads to 4 complete revolutions, $\tilde{\Omega}=10$ leads to 14.

Chapter 6

6.1 From (6.18) we have

$$\frac{d^2u}{d\theta^2}+u=\frac{cu}{m\hbar^2},$$

and from (6.15) we find that $\hbar=(r\dot{\theta})r=(c/md^2)^{1/2}d=(c/m)^{1/2}$. So $d^2u/d\theta^2=0$, and therefore

$$u=A\theta+B.$$

But (6.17) gives $\dot{r}=-\hbar\, du/d\theta=v$ at $\theta=0$, so $-\hbar A=v$, i.e. $A=-v/dv_c$. Moreover, $u=1/d$ when $\theta=0$, so $B=1/d$, and therefore

$$r=\frac{d}{1-\dfrac{v}{v_c}\theta}.$$

If $v<0$ the denominator becomes steadily larger as θ increases, so r becomes smaller and smaller, and the particle spirals in toward the origin. If $v>0$ the particle begins by spiralling out, but note that $r\to\infty$ as $\theta\to v_c/v$, so that it eventually asymptotes to a line radially outward from the origin.

6.2 The key result in each case is (6.18), i.e.

$$\frac{\mathrm{d}^2u}{\mathrm{d}\theta^2}+u=\frac{f\left(\dfrac{1}{u}\right)}{m\hbar^2u^2}.$$

(i) $r=\mathrm{e}^{-k\theta}$ so $u=1/r=\mathrm{e}^{k\theta}$. So

$$\frac{\mathrm{d}^2u}{\mathrm{d}\theta^2}+u=(k^2+1)\,\mathrm{e}^{k\theta}=(k^2+1)u,$$

and therefore $f(1/u)\propto u^3$, i.e. $f(r)\propto 1/r^3$.

(ii) Let the centre of the circular arc be at $r=a$, $\theta=0$. Then as the angle subtended by a diameter is a right angle, the polar equation of the arc is $r=2a\cos\theta$. So

$$u=\frac{1}{2a}\sec\theta,$$

$$\frac{\mathrm{d}u}{\mathrm{d}\theta}=\frac{1}{2a}\sec\theta\tan\theta,$$

$$\frac{\mathrm{d}^2u}{\mathrm{d}\theta^2}=\frac{1}{2a}(\sec\theta\tan^2\theta+\sec^3\theta),$$

and therefore

$$\frac{\mathrm{d}^2u}{\mathrm{d}\theta^2}+u=\frac{1}{2a}\cdot 2\sec^3\theta=8a^2u^3.$$

So $f(1/u)\propto u^5$, and therefore $f(r)\propto 1/r^5$.

6.3 By (6.32),

$$T=2\pi\left(\frac{a^3}{GM}\right)^{1/2},$$

and using (6.28) and (6.29) we may rewrite this as

$$T=2\pi\left(\frac{d^3}{GM}\right)^{1/2}\frac{1}{(2-v^2/v_c^2)^{3/2}}.$$

But (6.35) shows that the unit of time in the non-dimensionalization is $(d^3/GM)^{1/2}$, so

$$\tilde{T}=\frac{2\pi}{(2-v^2/v_c^2)^{3/2}}.$$

A straightforward way of incorporating this into the adapted NPHASE program is to insert directly after the INPUT statement

```
LOCATE 2, 1 : PRINT "period"
LOCATE 3, 1 : PRINT 2 * 3.14159/(2 - v ^ 2) ^ 1.5.
```

A crude comparison with the step-by-step method can be made simply by pressing the PAUSE key as the 'planet' completes one orbit.

6.4 The gravitational force on m_1 is Gm_1m_2/d^2, so by (6.4)

$$\frac{m_1v_1^2}{r_1} = \frac{Gm_1m_2}{d^2}.$$

On using the fact that $r_1 = (m_2/M)d$, as C is the centre of mass, we obtain (6.42a), and a similar argument for m_2 leads to (6.42b). The angular velocity of m_1, say, will be v_1/r_1, which is then seen to be $(GM/d^3)^{1/2}$.

6.5 To determine what happens next a variable step size is necessary with, say, hscale= .1. At $\tilde{t} = 5.0$, m_1 and m_2 form a pair, but this is broken up by the return of m_3 at $\tilde{t} \approx 10.0$, and a longer-lived m_1/m_3 pair is then formed. This can be confirmed by rerunning the program with hscale=0.05, say, and the outcomes can be compared directly on screen with the same simple modification used in the program NXT.

With $\tilde{m}_3 = 0.1$ instead, m_3 is at first abruptly expelled, more or less where it came from, but it returns at $\tilde{t} = 5$. The outcome of that return, and the whole subsequent motion, is changed completely if the initial y-coordinate of m_3 is 0.75075, rather than 0.75.

Chapter 7

7.1

(i)
$$\frac{\partial z}{\partial x} = 2x, \quad \frac{\partial z}{\partial t} = 0.$$

(ii) $z = x^2 - 2xct + c^2t^2$ so

$$\frac{\partial z}{\partial x} = 2x - 2ct = 2(x - ct),$$

$$\frac{\partial z}{\partial t} = -2xc + 2c^2t = -2c(x - ct).$$

Alternatively, write $z = X^2$ with $X = x - ct$ and use

$$\frac{\partial z}{\partial x} = \frac{\mathrm{d}z}{\mathrm{d}X}\frac{\partial X}{\partial x},$$

which is none other than the chain rule (2.9), because t is being held constant in the calculation of both $\partial z/\partial x$ and $\partial X/\partial x$, so z and X are being treated there effectively as functions of x only.

(iii) Write

$$z = \frac{1}{1+X^2} \qquad \text{with } X = x - ct.$$

$$\frac{\partial z}{\partial x} = \frac{\mathrm{d}z}{\mathrm{d}X}\frac{\partial X}{\partial x} = -\frac{2X}{(1+X^2)^2}, \quad \text{etc.}$$

7.2

$$\frac{\partial z}{\partial x} = f'(x)\sin\omega t, \qquad \frac{\partial z}{\partial t} = \omega f(x)\cos\omega t$$

$$\frac{\partial^2 z}{\partial x^2} = f''(x)\sin\omega t, \qquad \frac{\partial^2 z}{\partial t^2} = -\omega^2 f(x)\sin\omega t.$$

So the factors of $\sin\omega t$ cancel in (7.2) and we are left with an *ordinary* differential equation for $f(x)$:

$$T\frac{\mathrm{d}^2 f}{\mathrm{d}x^2} + \rho\omega^2 f = 0.$$

This is linear, with constant coefficients, so by (3.27) and (3.28)

$$f(x) = C\cos\left(\frac{\rho}{T}\right)^{1/2}\omega x + D\sin\left(\frac{\rho}{T}\right)^{1/2}\omega x$$

where C and D are arbitrary constants. But the end-points of the string are fixed, so $f(0) = f(l) = 0$. The first of these gives $C = 0$, and the second then gives either $D = 0$ (in which case $f = 0$ and the string is not vibrating at all) *or*

$$\sin\left(\frac{\rho}{T}\right)^{1/2}\omega l = 0,$$

so that $(\rho/T)^{1/2}\omega l = N\pi$, where N is an integer. The corresponding displacement is

$$z = D\sin\frac{N\pi x}{l}\sin\omega t,$$

and this is zero for all t wherever $\sin(N\pi x/l)$ is zero, namely at

$$x = \frac{Ml}{N}, \quad M = 1, 2, \ldots, N-1,$$

together, of course, with the end-points $x = 0, l$.

7.3 (i) At any particular time t, the maximum value of T is clearly at $x = 0$, and this maximum value is $T_0/(1 + 4\kappa t/a^2)^{1/2}$, which decreases with time.

But the *spreading* is due essentially to the factor $(a^2+4\kappa t)^{-1}$ in the exponent.

To see this, note that at any particular time, T falls to (say) $\frac{1}{10}$ of its maximum value at that time at $|x|=d$, where $d^2=(a^2+4\kappa t)\log 10$. So $2d$, which serves as a rough indicator of the 'width' of the hump, increases with time, eventually in proportion to $t^{1/2}$.

(ii) Let $\phi=-x^2/4\kappa t$.

$$T=\frac{1}{t^{1/2}}\,\mathrm{e}^{\phi}$$

$$\frac{\partial T}{\partial t}=-\frac{1}{2t^{3/2}}\,\mathrm{e}^{\phi}+\frac{1}{t^{1/2}}\,\frac{\partial \phi}{\partial t}\,\mathrm{e}^{\phi}$$

$$=\left(-\frac{1}{2t^{3/2}}+\frac{1}{t^{1/2}}\cdot\frac{x^2}{4\kappa t^2}\right)\mathrm{e}^{\phi}$$

$$\frac{\partial T}{\partial x}=\frac{1}{t^{1/2}}\,\frac{\partial \phi}{\partial x}\,\mathrm{e}^{\phi}=\frac{-x}{t^{1/2}2\kappa t}\,\mathrm{e}^{\phi}$$

$$\frac{\partial^2 T}{\partial x^2}=-\frac{1}{2\kappa t^{3/2}}\,\mathrm{e}^{\phi}-\frac{x}{2\kappa t^{3/2}}\left(\frac{-x}{2\kappa t}\right)\mathrm{e}^{\phi}$$

$$=\frac{1}{\kappa}\left(-\frac{1}{2t^{3/2}}+\frac{x^2}{4\kappa t^{5/2}}\right)\mathrm{e}^{\phi}=\frac{1}{\kappa}\,\frac{\partial T}{\partial t}.$$

7.5 $\phi(x,y)=xy+y^2$, so

$$\frac{\partial \phi}{\partial x}=y,\qquad \frac{\partial \phi}{\partial y}=x+2y.$$

Now, $x=t$ and $y=t^2$, so

$$\frac{\partial \phi}{\partial x}\,\frac{\mathrm{d}x}{\mathrm{d}t}+\frac{\partial \phi}{\partial y}\,\frac{\mathrm{d}y}{\mathrm{d}t}=y\cdot 1+(x+2y)\cdot 2t$$

$$=t^2+(t+2t^2)2t$$

$$=3t^2+4t^3.$$

But when ϕ is written directly as a function of t alone, we have $\phi=t^3+t^4$, so

$$\frac{\mathrm{d}\phi}{\mathrm{d}t}=3t^2+4t^3,$$

and the result is seen to be true, at least in this particular case.

Chapter 8

8.1 With reference to Fig. 8.8, the time taken is

$$T = \frac{(x^2 + d_1^2)^{1/2}}{c_1} + \frac{\left[(L-x)^2 + d_2^2\right]^{1/2}}{c_2},$$

so

$$\frac{\mathrm{d}T}{\mathrm{d}x} = \frac{x}{c_1(x^2 + d_1^2)^{1/2}} - \frac{(L-x)}{c_2\left[(L-x)^2 + d_2^2\right]^{1/2}}$$
$$= \frac{\sin\theta_1}{c_1} - \frac{\sin\theta_2}{c_2}.$$

Setting this equal to zero gives the result, and the stationary point is a minimum because

$$\frac{\mathrm{d}^2T}{\mathrm{d}x^2} = \frac{d_1^2}{c_1(x^2 + d_1^2)^{3/2}} + \frac{d_2^2}{c_2\left[(x-L)^2 + d_2^2\right]^{3/2}} > 0.$$

8.2 $$\frac{\mathrm{d}}{\mathrm{d}x}\left\{\frac{y\dot{y}}{(1+\dot{y}^2)^{1/2}}\right\} - (1+\dot{y}^2)^{1/2}$$
$$= \frac{(1+\dot{y}^2)^{1/2}(\dot{y}^2 + y\ddot{y}) - y\dot{y}\tfrac{1}{2}(1+\dot{y}^2)^{-1/2}2\dot{y}\ddot{y} - (1+\dot{y}^2)^{3/2}}{(1+\dot{y}^2)}.$$

This being zero, we have

$$(1+\dot{y}^2)(\dot{y}^2 + y\ddot{y}) - y\dot{y}^2\ddot{y} - (1+\dot{y}^2)^2 = 0,$$

which simplifies to

$$y\ddot{y} - \dot{y}^2 = 1.$$

To solve this equation, write

$$\dot{y} = v,$$
$$\dot{v} = \frac{1+v^2}{y},$$

so

$$\frac{\mathrm{d}v}{\mathrm{d}y} = \frac{1+v^2}{vy},$$
$$\int\frac{v\,\mathrm{d}v}{1+v^2} = \int\frac{\mathrm{d}y}{y},$$
$$\tfrac{1}{2}\log(1+v^2) = \log y + k,$$
$$1 + v^2 = Ay^2.$$

So

$$\dot{y} = \pm(Ay^2 - 1)^{1/2},$$

$$\int \frac{\mathrm{d}y}{(Ay^2 - 1)^{1/2}} = \pm x.$$

Make the substitution $y = A^{-1/2}\cosh p$, because

$$(\cosh^2 p - 1)^{1/2} = \sinh p \qquad \text{and} \qquad \frac{\mathrm{d}y}{\mathrm{d}p} = \frac{1}{\sqrt{A}} \sinh p,$$

so

$$\frac{p}{\sqrt{A}} = \pm x + d,$$

and therefore writing $c = 1/\sqrt{A}$

$$y = c\cosh\left(\pm\frac{x}{c} + l\right).$$

But the boundary conditions are $y = 1$ at $x = \pm a$, so it follows that $l = 0$ and

$$c\cosh\left(\frac{a}{c}\right) = 1,$$

the solution being

$$y = c\cosh\left(\frac{x}{c}\right).$$

To determine when c is real, put $\xi = a/c$ and consider the equation

$$\cosh\xi = \frac{\xi}{a}, \qquad a > 0.$$

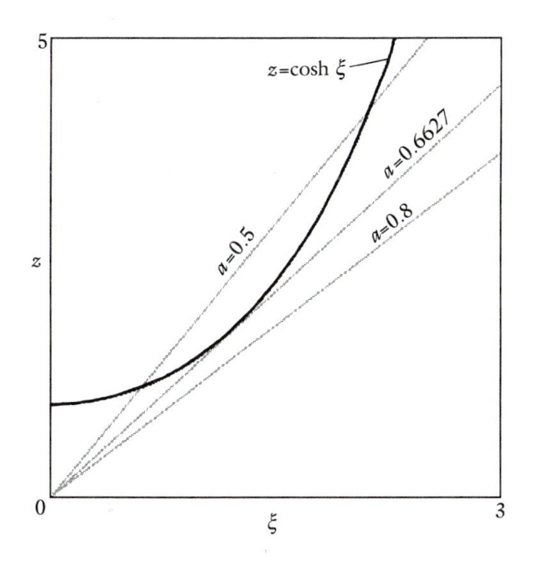

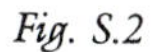
Fig. S.2

We see from Fig. S.2 that the curves $z = \cosh \xi$ and $z = \xi/a$ will not intersect if a is too large, and the critical case is when they touch, so that the values of z *and* $\mathrm{d}z/\mathrm{d}\xi$ coincide:

$$\cosh \xi = \xi/a \qquad \text{and} \qquad \sinh \xi = 1/a.$$

In this critical case, then, $\coth \xi = \xi$, and the solution of this is found numerically to be $\xi = 1.1997$. The corresponding value of a, namely $(\sinh \xi)^{-1}$, is 0.6627.

For smaller values of a we see from Fig. S.2 that there are in fact *two* values of ξ at which the curves intersect, but it turns out that only the smaller of these gives a surface area for the film which is a local *minimum* with respect to small variations in that surface.

8.3 (i) $T = \frac{1}{2}m(l\dot{\theta})^2$, $V = mgl(1 - \cos\theta)$, so

$$L = \tfrac{1}{2}ml^2\dot{\theta}^2 - mgl(1 - \cos\theta).$$

Equation (8.20) yields

$$\frac{\mathrm{d}}{\mathrm{d}t}\left(\frac{\partial L}{\partial \dot{\theta}}\right) - \frac{\partial L}{\partial \theta} = 0,$$

and as

$$\frac{\partial L}{\partial \dot{\theta}} = ml^2\dot{\theta}, \qquad \frac{\partial L}{\partial \theta} = -mgl \sin\theta,$$

we have

$$ml^2\ddot{\theta} + mgl\sin\theta = 0,$$

which simplifies to (5.2).

(ii) The Lagrangian is

$$L = \tfrac{1}{2}m\dot{x}_1^2 + \tfrac{1}{2}m\dot{x}_2^2 - \left[\tfrac{1}{2}\alpha x_1^2 + \tfrac{1}{2}\alpha(x_2 - x_1)^2 + \tfrac{1}{2}\alpha x_2^2\right],$$

where we have used the expression $\frac{1}{2}\alpha(\text{extension})^2$ for the elastic potential energy in each of the three springs (see Ex. 3.6).

Equation (8.20) yields the two equations

$$\frac{\mathrm{d}}{\mathrm{d}t}\left(\frac{\partial L}{\partial \dot{x}_1}\right) - \frac{\partial L}{\partial x_1} = 0,$$

$$\frac{\mathrm{d}}{\mathrm{d}t}\left(\frac{\partial L}{\partial \dot{x}_2}\right) - \frac{\partial L}{\partial x_2} = 0,$$

so

$$m\ddot{x}_1 = -\alpha x_1 + \alpha(x_2 - x_1),$$

$$m\ddot{x}_2 = -\alpha(x_2 - x_1) - \alpha x_2,$$

in agreement with (5.19a, b).

Chapter 9

9.1
$$u = \frac{\partial x}{\partial t} \quad \text{holding } X, Y, Z \text{ all constant}$$
$$= -\Omega X \sin \Omega t - \Omega Y \cos \Omega t = -\Omega y.$$

Similarly $v = \Omega x$, $w = 0$.

An incompressible fluid *could* flow like this, as

$$\frac{\partial u}{\partial x} + \frac{\partial v}{\partial y} + \frac{\partial w}{\partial z} = 0$$

(see (9.4)), each individual term here being zero in this particular case.

$$x^2 + y^2 = (X \cos \Omega t - Y \sin \Omega t)^2 + (Y \cos \Omega t + X \sin \Omega t)^2$$
$$= X^2 + Y^2,$$

so each fluid particle moves in such a way that it stays at a constant distance from the $x = 0$, $y = 0$ axis. The flow as a whole is in fact simple uniform rotation, with constant angular velocity Ω.

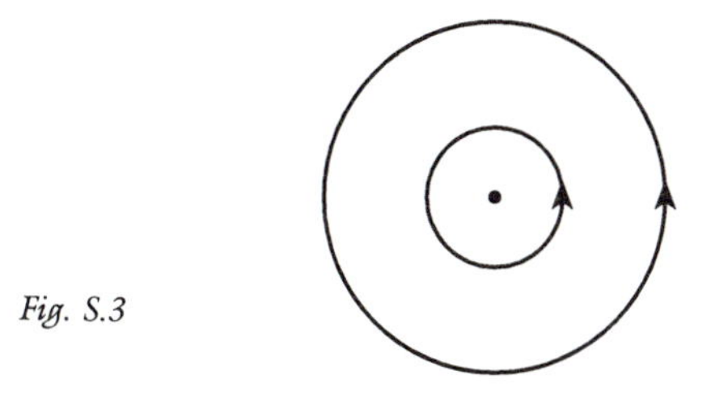

Fig. S.3

9.2 With $u = \alpha x$, $v = -\alpha y$, $w = 0$ and α constant we have

$$\frac{\partial u}{\partial x} = \alpha, \quad \frac{\partial u}{\partial y} = 0, \quad \frac{\partial^2 u}{\partial x^2} = 0, \quad \frac{\partial^2 u}{\partial y^2} = 0,$$

$$\frac{\partial v}{\partial x} = 0, \quad \frac{\partial v}{\partial y} = -\alpha, \quad \frac{\partial^2 v}{\partial x^2} = 0, \quad \frac{\partial^2 v}{\partial y^2} = 0$$

so (9.9c) is certainly satisfied, while (9.9a) and (9.9b) reduce to

$$\rho \alpha^2 x = -\frac{\partial p}{\partial x},$$

$$\rho \alpha^2 y = -\frac{\partial p}{\partial y}.$$

At this point the serious question arises of whether there *exists* a function $p(x, y)$ which satisfies *both* of these equations; if not, then the original flow cannot be a solution of the viscous flow equations. Happily, such a function does exist in this case:

$$p = c - \tfrac{1}{2}\rho\alpha^2(x^2 + y^2)$$

satisfies both equations, for any constant c.

9.3 To solve equation (9.15) we may *either*

(i) write $v = \mathrm{d}u/\mathrm{d}x$, obtaining

$$\varepsilon \frac{\mathrm{d}v}{\mathrm{d}x} + v = 1,$$

which is first order, linear and soluble by the method of integrating factors (Section 3.2), *or*

(ii) treat it as a second-order linear equation with constant coefficients (Section 3.4).

By either method, the general solution of the equation is

$$u = A + B\,\mathrm{e}^{-x/\varepsilon} + x,$$

and the constants A and B are then determined by the boundary conditions $u(0) = 0$, $u(1) = 2$.

If $\varepsilon < 0$ write the solution as

$$u = x + \frac{1 - \mathrm{e}^{x/|\varepsilon|}}{1 - \mathrm{e}^{1/|\varepsilon|}}.$$

Now, $1/|\varepsilon|$ is large, so $\mathrm{e}^{1/|\varepsilon|}$ is enormously large and the 1 in the denominator may be neglected giving

$$u \doteqdot x - \mathrm{e}^{-1/|\varepsilon|} + \mathrm{e}^{(x-1)/|\varepsilon|}.$$

The second term here is, again, tiny, and so is the last term for virtually all x in the interval $0 \le x \le 1$ *except when x is very close to* 1. So

$$u \doteqdot x$$

here, except when x is nearly equal to 1, when the last term rapidly changes from being negligibly small to precisely 1 at $x = 1$.

In place of Fig. 9.8 we therefore have Fig. S.4, with the 'boundary layer' occurring on the other boundary.

Interestingly, then, the solution to the problem (9.15), (9.16) as $|\varepsilon| \to 0$ is quite different, depending on whether ε tends to zero through positive or through negative values. [We can, of course, take some mystery out of this by

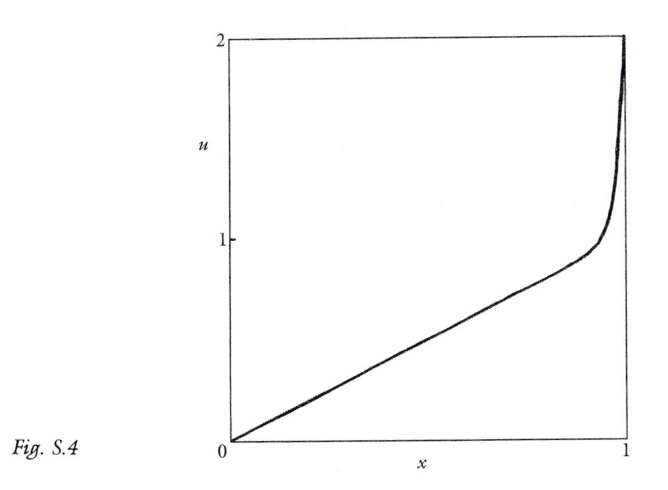

Fig. S.4

noting that if ε is actually *equal* to zero, the problem (9.15), (9.16) has *no solution at all*.]

9.4 The general solution is

$$u = A\,\mathrm{e}^{mx} + B\,\mathrm{e}^{-mx} + 1,$$

where

$$m = \frac{1}{\varepsilon^{1/2}},$$

and on choosing A and B to satisfy the boundary conditions we find

$$u = 1 + \frac{(\mathrm{e}^{-m} - 1)\,\mathrm{e}^{mx} + (1 - \mathrm{e}^{m})\,\mathrm{e}^{-mx}}{\mathrm{e}^{m} - \mathrm{e}^{-m}}.$$

If $\varepsilon \ll 1$ then $m \gg 1$, and so

$$u \doteqdot 1 + \frac{-\mathrm{e}^{mx} - \mathrm{e}^{m(1-x)}}{\mathrm{e}^{m}},$$

i.e.

$$u \doteqdot 1 - \mathrm{e}^{-m(1-x)} - \mathrm{e}^{-mx}.$$

The last two terms are both negligibly small for most x in the range $0 \leq x \leq 1$, but the second is significant very close to $x = 1$, and the third is significant very close to $x = 0$. Here, then, there are thin boundary layers on *both* boundaries (Fig. S.5).

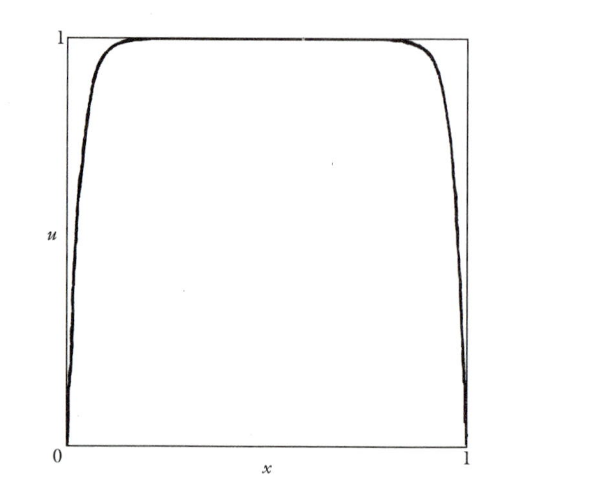

Fig. S.5

Chapter 10

10.1 Last part: it settles on the *left*-hand equilibrium, with $\theta_0 < 0$, after being 'bounced' out of the region $\theta > 0$.

10.2 For circular motion about the origin $\ddot{r} = 0$ so $h^2 = c/ma^{n-3}$. But h is defined as $r^2\dot{\theta}$, and for the given circular motion this is $a^2\Omega$, so $a^4\Omega^2 = c/ma^{n-3}$.

In view of these results we may rewrite the differential equation as

$$\ddot{\eta} - \frac{c}{ma^n}\left(\frac{a}{r}\right)^3 = -\frac{c}{ma^n}\left(\frac{a}{r}\right)^n,$$

i.e. as

$$\ddot{\eta} - \frac{c}{ma^n}\left(1 + \frac{\eta}{a}\right)^{-3} = -\frac{c}{ma^n}\left(1 + \frac{\eta}{a}\right)^{-n}.$$

As yet we have made no approximation, but we now assume η/a to be small, and expand binomially correct to first order in η/a (see (2.16)) to obtain the linearized equation

$$\ddot{\eta} - \frac{c}{ma^n}\left(1 - \frac{3\eta}{a}\right) = -\frac{c}{ma^n}\left(1 - \frac{n\eta}{a}\right),$$

i.e.

$$\ddot{\eta} + (3 - n)\frac{c}{ma^{n+1}}\,\eta = 0.$$

We therefore have simple harmonic oscillations for η if $n < 3$ but instability of the $\eta = 0$ state if $n > 3$.

10.3 Write (10.22) as

$$\ddot{\theta} = \frac{1}{M}[\varepsilon - F(\theta)],$$

then write $\theta = \theta_0 + \theta_1(t)$, so that to first order in θ_1 (assume small)

$$\ddot{\theta}_1 = \frac{1}{M}[\varepsilon - F(\theta_0) - \theta_1 F'(\theta_0)],$$

by Taylor's theorem. The first two terms on the right-hand side cancel, by the equilibrium condition $\varepsilon = F(\theta_0)$, so

$$\ddot{\theta}_1 = -\frac{1}{M}F'(\theta_0)\cdot\theta_1.$$

The equilibrium is therefore stable if $F'(\theta_0) > 0$ and unstable if $F'(\theta_0) < 0$. This implies that the equilibria in Fig. 10.12(a) are stable, because the graph of ε against θ_0 has positive slope. Similarly, in Fig. 10.12(b) the equilibria are stable where $\mathrm{d}\varepsilon/\mathrm{d}\theta_0 > 0$ but unstable in the middle portion, where $\mathrm{d}\varepsilon/\mathrm{d}\theta_0 < 0$. [The associated claims made in respect of Fig. 10.13 then follow at once.]

The jumps occur when ε increases/decreases past the critical values corresponding to the local maximum/minimum in Fig. 10.12(b); these occur at $\theta_0 = \mp 2^{1/2}(1 - M^{-1})^{1/2}$, as noted in the text, and the corresponding values of ε then come directly from (10.23).

10.4 A constant solution $\theta = \theta_0$ must satisfy

$$\sin\theta_0 = \tilde{\Gamma},$$

and this can only happen if $\tilde{\Gamma} < 1$. For each value of $\tilde{\Gamma}$ in the range $0 < \tilde{\Gamma} < 1$ there are then two possibilities, one with $\theta_0 < \pi/2$ and $\cos\theta_0 > 0$, the other with $\pi/2 < \theta_0 < \pi$ and $\cos\theta_0 < 0$.

To explore stability, write $\theta = \theta_0 + \theta_1(\mathrm{t})$, and expand $\sin\theta$ in a Taylor series about θ_0 correct to first order in θ_1, assumed small, so that

$$\ddot{\theta}_1 + \tilde{k}\dot{\theta}_1 + \sin\theta_0 + \cos\theta_0\cdot\theta_1 = \tilde{\Gamma},$$

so

$$\ddot{\theta}_1 + \tilde{k}\dot{\theta}_1 + \cos\theta_0\cdot\theta_1 = 0.$$

If damping is absent, so that $\tilde{k} = 0$, we see at once that the equilibrium with $\cos\theta_0 > 0$ is stable while the other equilibrium is unstable.

But damping does not, in any case, affect this conclusion, because by (3.32) the general solution is then

$$\theta_1 = E\,e^{m_1\tilde{t}} + F\,e^{m_2\tilde{t}},$$

where

$$m_1, m_2 = -\frac{\tilde{k}}{2} \pm \left[\frac{\tilde{k}^2}{4} - \cos\theta_0\right]^{1/2}.$$

If $\cos\theta_0 < 0$ it follows at once that *one* of m_1, m_2 will be real and positive, giving exponential growth of θ_1 with time and hence instability of the equilibrium state in question. If, however, $\cos\theta_0 > 0$, then either (a) the contents of the brackets are negative, in which case we have decaying oscillations about $\theta = \theta_0$ and therefore stability (cf. Ex. 5.1), or (b) the square root is real, but less than $\tilde{k}/2$, so that both m_1 and m_2 are real and negative, giving straightforward non-oscillatory decay of the disturbance θ_1 and, again, stability of the $\theta = \theta_0$ state.

Chapter 11

11.1 Suggested changes to `NXT`:

```
n = 2
xm = 3 : tm = 180#
```

Replace `k = .05# : w = 1# : a = 7.5#` by `eps = .1#`
Replace `t = 0# : xc(3) = 0#` by `t = 0#`

```
f(2) =-eps * (x(1) ^ 2# - 1#) * x(2) - x(1)
```

Delete `f(3) = 1#`

On running the program, `INPUT`, say `.01, 0, .1, 10`, and confirm by halving the step size, with an `INPUT .01, 0, .05, 12`, say.

For larger `eps`, setting `tm=50#`, try `eps=10#` and then `eps=20#`. The time steps `h` then have to be taken significantly smaller in order to resolve the rapidly changing part of the oscillation correctly and get consistent results. Nonetheless, even with `eps=50#` I obtain 82.5 as the period, originally using `h=.005` and confirmed using `h=.0025`. This compares with a value of 80.7 if we use the asymptotic formula 1.614ε, which is formally valid in the limit $\varepsilon \to \infty$.

11.2 Direct substitution of $x = r\cos\theta$, $y = r\sin\theta$ gives

$$\dot{r}\cos\theta - r\sin\theta\cdot\dot{\theta} = r\sin\theta + \varepsilon r(1 - r^2)\cos\theta,$$
$$\dot{r}\sin\theta + r\cos\theta\cdot\dot{\theta} = -r\cos\theta + \varepsilon r(1 - r^2)\sin\theta.$$

Multiplying the first by $\cos\theta$, the second by $\sin\theta$, and adding gives

$$\dot{r} = \varepsilon r(1 - r^2),$$

and a similar elimination of $\dot{r}$ gives

$$\dot{\theta} = -1.$$

The first equation is separable, and can then be split into partial fractions, so that

$$\int \frac{1}{r} + \frac{\frac{1}{2}}{1-r} - \frac{\frac{1}{2}}{1+r}\,\mathrm{d}r = \int \varepsilon\,\mathrm{d}t.$$

So

$$\log \frac{r}{|1 - r^2|^{1/2}} = \varepsilon t + \text{constant},$$

while the other equation integrates immediately to give $\theta = \text{constant} - t$.

Let $r = r_0$ and $\theta = \theta_0$ at $t = 0$. On determining the two constants and rearranging we have

$$r = \frac{1}{\left[1 + \left(\dfrac{1}{r_0^2} - 1\right)\mathrm{e}^{-2\varepsilon t}\right]^{1/2}}, \qquad \theta = \theta_0 - t.$$

As $t \to \infty$, $\mathrm{e}^{-2\varepsilon t} \to 0$, so $r \to 1$, and any phase point therefore moves closer and closer to the unit circle in the phase plane as it rotates about the origin with angular velocity $\dot{\theta} = -1$.

In contrast to the van der Pol equation, the final limit-cycle oscillation is independent of ε; it is only the rate at which it is approached that is controlled by ε. It is interesting to use `2PHASE` to contrast the cases $\varepsilon = 0.1$ and $\varepsilon = 3$, for example.

11.3 No. The non-autonomous system $\dot{x} = f(x, t)$ can be recast as the *autonomous* two-dimensional system

$$\dot{x} = f(x, t),$$
$$\dot{t} = 1,$$

and no chaos is possible in such a system, by Section 11.3.

There are, in fact, *two* first integrals of the Euler equations. The more obvious, perhaps, is obtained by multiplying the equations by x, y, z, respectively, and adding, to obtain

$$Ax\dot{x} + By\dot{y} + Cz\dot{z} = 0$$

and therefore

$$Ax^2 + By^2 + Cz^2 = \text{constant}.$$

But multiplying by Ax, By, Cz and adding also does the trick, for the right-hand contributions again all cancel and as a consequence

$$A^2x^2 + B^2y^2 + C^2z^2 = \text{constant}.$$

The phase paths themselves are therefore given by the intersections of these two sets of surfaces.

The existence of even just one first integral $F(x, y, z) = \text{constant}$ here precludes chaos. To see this, either (i) note that a phase point is then confined to a particular *surface* in three-dimensional phase space, and is therefore effectively in a two-dimensional space, or (ii) note that we could use $F(x, y, z) = \text{constant}$ to eliminate, say, z from the first two equations, and those two would then become an autonomous system of only *second* order, so that x and y (and therefore z) could not evolve chaotically, by Section 11.3.

11.4 On running `NSENSIT`, as written, tiny differences in the resulting phase paths are perceptible by $t \sim 9$. These are quite substantial by $t = 10$, and the four phase points have become completely scattered over the attractor by $t = 12$. Confirm that the numerical integrations are, apparently, *just* trustworthy over this time interval, by repeating the computation with $h = .0015$ rather than $h = .003$.

To use `NVARY` to explore the Rössler equations, change `Equations` to

```
f(1) = -x(2) - x(3)
f(2) = x(1) + .2# * x(2)
f(3) = .2# + (x(1) - c) * x(3)
```

change the parameter `w` occurring at several points in the program, as written, to `c`, and delete the line `k=.1# : b=.04#`. Then `INPUT`, say, `c=2.5` and gradually increase `c` by pressing `q` at suitable intervals and making the desired adjustment, as explained in the notes for the program `NVARY`.

The system behaviour for `4.5 < c < 6` is mostly chaotic, but a fairly wide 'periodic window' exists around `c=5.3`, where there is an oscillation which repeats itself after three cycles. This period-3 oscillation undergoes a period-doubling cascade, starting at `c` ≑ `5.37`. There is also a more narrow period-5 window, with similar structure, around `c` ≑ `4.7`.

11.5 A fairly small h is necessary to confirm *both* solutions in Fig. 11.1; I obtain consistent results for each, over the time interval in question, with $h = 0.05$, 0.025, or 0.0125.

A step size $h = 0.1$ should be sufficient, however, for the exploration of limit cycles with `NXTWAIT` which follows. Change the second line of 'Setting up graphics' to, say,

```
xm = 2 : twait = 100# : tm = 100#
```

and, of course, change the parameters to

```
k = .08# : w = 1# : a = .2#
```

With `x2=0` initially the results of different initial values of `x1` should be as follows:

1	period 2π, larger amplitude than the others
0.2	period 2π
−0.9	period 4π, asymmetric motion about $x=0$
−0.7	period 4π, asymmetric motion about $x=0$
0	period 6π.

11.6 One suitable program would be

```
DEFDBL A, X
a = 3.2#
x = .1#
FOR n = 1 TO 500
 x = a * x * (1# - x)
 PRINT n, x
NEXT
```

where a plays the role of λ.

Figure 11.17 was in fact produced with the closely related program

```
DEFDBL A, X

 CLS : SCREEN 9
 VIEW (10, 10) - (550, 300), 0, 9
 WINDOW (2.5, 0) - (4, 1)

FOR a = 2.5# TO 4# STEP .001#

 x = .1#

  FOR n = 1 TO 1000
   x = a * x * (1# - x)
    IF n > 900 THEN
      PSET (a, x)
    END IF
  NEXT

NEXT
```

Chapter 12

12.1 On running VIBRAPEN try inputting, say, $\tilde{a} = 0.2$, $\tilde{\omega} = 10$, with θ (i.e. angle) as 3.1 and $\dot{\theta}$ (i.e. angvel) as 0 at $\tilde{t} = 0$. It is worth confirming that the inverted state is *not* stable with, say (i) $\tilde{a} = 0.2$, $\tilde{\omega} = 7$, (ii) $\tilde{a} = 0.1$, $\tilde{\omega} = 10$, (iii) $\tilde{a} = 0.6$, $\tilde{\omega} = 10$.

For the dancing oscillations of Fig. 12.3, try the initial conditions in the caption, but note that for substantially smaller initial values of $|\dot{\theta}|$ the pendulum simply reverts to the classical inverted state shown in Fig. 12.1.

12.2 Construct the diagram below, with R the mid-point of QP, and note that if we let α denote the angle which QP makes with the horizontal, then, $\angle QOR = \alpha$. Let $QP = d = 2l \sin \alpha$. Then equating the gain in kinetic energy to the loss in potential energy gives

$$\begin{aligned} \tfrac{1}{2}mv^2 &= mg \cdot d \sin \alpha \\ &= mg \cdot \frac{d^2}{2l}. \end{aligned}$$

As l is constant, it follows that $v \propto d$.

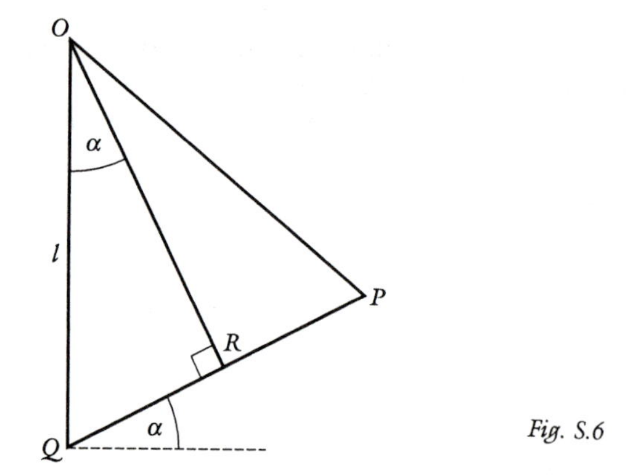

Fig. S.6

12.3 On running VIBRAPEN with $\tilde{a} = 0.1$ and $\tilde{\omega} = 2$ INPUT, say, $\theta = 0.01$ and $\dot{\theta} = 0$. The instability is then evident, and the pendulum settles into a regular, large-amplitude swinging oscillation by $\tilde{t} \sim 150$. This does not happen if $\tilde{\omega}$ is, say, 1.7 or 2.3 instead, though it would do if the drive amplitude $\tilde{a}$ were larger.

The answer to the *double* pendulum question is: yes, provided the damping is not too strong for the mode in question. Using (12.4) with $m = 0.5$, the relevant pivot vibration frequencies $\tilde{\omega}$ are 1.550 or 3.696. With $\tilde{a} = 0.1$ and $\tilde{k} = 0.1$ the second of these excites the fast mode—if we `INPUT`, say, $\theta_1 = 0.01$, $\theta_2 = 0.01$ at $\tilde{t} = 0$—but the first does not excite the slow mode. Yet if we reduce $\tilde{k}$ in the program to, say, 0.03, both modes are excited by the appropriate pivot vibration.

12.4 With $\tilde{a} = 0.1$, some suitable initial conditions are $\theta_1 = \theta_2 = 0.1$, $\dot{\theta}_1 = \dot{\theta}_2 = 0$ (for the downward-hanging state) or $\theta_1 = 1$, $\theta_2 = -1$, $\dot{\theta}_1 = \dot{\theta}_2 = 0$ (for the swinging oscillation).

With $\tilde{a} = 0.25$, some suitable initial conditions are $\theta_1 = 1.5$, $\theta_2 = -1.5$, $\dot{\theta}_1 = \dot{\theta}_2 = 0$ (swinging oscillation) or $\theta_1 = \theta_2 = 1$, $\dot{\theta}_1 = \dot{\theta}_2 = 5$ (whirling motion).

12.5 Try, say, $\theta_1 = 3.1$, $\theta_2 = 3.2$ first. On using (12.4) the inequalities (12.16) become

$$\tilde{a} < 0.45(1 - \sqrt{m}), \qquad \tilde{a}\tilde{\omega} > \sqrt{2}\,(1 + \sqrt{m})^{1/2},$$

where we have used the definitions (12.12). If we put $m = 0.5$ in these we obtain

$$\tilde{a} < 0.13, \qquad \tilde{a}\tilde{\omega} > 1.85$$

as criteria for stability of the inverted state, and $\tilde{a} = 0.1$, $\tilde{\omega} = 25$ satisfy these.

Because the drive frequency is quite large, the integration time step h should be taken quite small to obtain consistent results, say $h = 0.01$ or 0.02.

The inverted state is remarkably stable to disturbances in which the pendulums are reasonably well aligned with one another, but less so if θ_1 and θ_2 are at first very different, or if the two pendulums are initially on opposite sides of the upward vertical.

With the given value of $\tilde{\omega} = 25$, try $\tilde{a} = 0.04$ or $\tilde{a} = 0.16$ to see the two different types of collapse shown in Fig. 12.13. For upside-down dancing oscillations with $\tilde{a} = 0.11$ try $\theta_1 = 3.05$, $\theta_2 = 3.25$, $\dot{\theta}_1 = \dot{\theta}_2 = 0$ at $\tilde{t} = 0$.

Index